T0310174

COMMUNICATION PATTERNS OF ENGINEERS

COMMUNICATION PATTERNS OF ENGINEERS

CAROL TENOPIR
DONALD W. KING

IEEE Education Society, *Sponsor*

IEEE Professional Communication Society, *Sponsor*

IEEE PRESS

A JOHN WILEY & SONS, INC., PUBLICATION

For general information on our other products and services please contact our Customer Care Department within the U.S. at 877-762-2974, outside the U.S. at 317-572-3993 or fax 317-572-4002.

Wiley also publishes its books in a variety of electronic formats. Some content that appears in print, however, may not be available in electronic format.

Library of Congress Cataloging-in-Publication Data:

Tenopir, Carol.
    Communication patterns of engineers / Carol Tenopir, Donald W. King.
        p. cm.
    Includes bibliographical references and index.
    ISBN 0-471-48492-X (cloth : alk. paper)
        1. Communication in engineering.   I. King, Donald Ward, 1932–   II. Title.

TA158.5.T46 2004
620'.001'4—dc22   2003062037

Printed in the United States of America.

10 9 8 7 6 5 4 3 2

CONTENTS

PREFACE

In the fall of 2000 the Engineering Information Foundation (EIF) Board of Directors asked Donald W. King to advise them on possible future research directions. In a presentation to the Board, Mr. King recommended a four-phase approach to a research agenda regarding the communication of engineers, starting with a review of recent literature to identify where benchmark data exist, where there are gaps in research, and where future research would be beneficial to the engineering communities. As a result of this recommendation, the Board awarded a research grant to Carol Tenopir and Donald W. King to conduct a literature review and to present recommendations for future research directions. The report to EIF was the genesis of this book. The report focused on the literature from 1994 to the present pertaining to how engineers communicate. This book expands that focus to include literature from the 1960s to the present. The emphasis, however, remains on how engineers communicate, whether communication patterns have changed, and what might be done to improve communication of engineers.

This book broadly defines communication as encompassing information inputs such as seeking, locating, obtaining, and using information on the one hand and information outputs such as writing and oral communications. A particular emphasis is on how communication can be improved through education. We analyzed the literature that touches on these topics, particularly the research literature with engineers as the subject, either wholly or in part. We also extracted survey responses from engineers who were observed nearly every year from 1977 to 2003. These data provided useful insights into engineers' communication patterns and useful comparisons with science and other fields.

This project has been a group effort (not unlike the trend in engineering toward collaborative works). Project leaders Carol Tenopir and Donald King were ably assisted in all aspects by Rhyn Davies, Christine L. Ferguson, Edward Gray, and Scott Rice, graduate students at the University of Tennessee, School of Information Sciences. In addition, graduate students Katie Darraj, Keri-Lynn Paulson, Emily Urban, and Mercy Ebuen assisted with occasional specific tasks. At the University of Pittsburgh, School of Information Sciences, Sarah Aerni, Richard Daddieco, Matt Herbison, and Gina Cecchetti also made helpful contributions.

Without the initial funding and encouragement by the Engineering Information Foundation, this project would not have been possible. We would like to extend special thanks to the EIF Board, including; Melvin Day; Thomas R. Buckman, President; Anne M. Buck, Vice-President; Hans Rütimann, Secretary; John J. Regazzi, Director; Julie A. Shimar, Director; and Ruth A. Miller, Executive Associate. Their comments, corrections, and encouragement were essential to this book. The authors also wish to partially dedicate this book to the memory of Anne M. Buck, a dedicated and caring engineering librarian.

CAROL TENOPIR
DONALD W. KING

Knoxville, Tennessee
Pittsburgh, Pennsylvania
August 2003

COMMUNICATION PATTERNS OF ENGINEERS

1

INTRODUCTION

1.1 FOCUS OF THE BOOK

This book is a review and analysis of the literature and presentation of data from a series of surveys that attempts to provide insights into how engineers communicate. Much of the focus of the book is on the professional aspects of engineers' work, the information resources used to perform their work, and information output from their work that is communicated to others. Many of our studies and those of others dealt with traditional interpersonal and written communication channels. Together, these studies provide abundant evidence of the many factors that motivate engineers to use various communication channels. However, it seems clear that new technologies, such as the World Wide Web and electronic publishing, are having a profound effect on engineering communication patterns. We believe that knowledge and understanding of engineers' motives, incentives, and reasons for communicating in the past will help frame future communication practices.

During the late 1980s and early 1990s, the Internet, and specifically the World Wide Web, became popular, making electronic and digitally based products (i.e., electronic journals) not only possible, but economically practical. By the late 1990s, electronic products became widely available and accepted by authors

Communication Patterns of Engineers. By Carol Tenopir and Donald W. King **1**
ISBN 0-471-48492-X © 2004 Institute of Electrical and Electronics Engineers

and readers. The Internet has dramatically increased the potential for both informal and formal communication. People have the option of easy and immediate contact with friends and colleagues all over the world, there has been an evolution of intensive groups of engineers on the Internet with interests in materials, nanotech, electronics, and so on. They can choose which format of many best suits their communication and information needs and requirements. Libraries also now have the option of choosing between print or electronic formats. Libraries and information centers exist to provide information services to their users, so it is important to find out which formats users prefer and how potential benefits offered by electronic resources will facilitate the research and development process and help (or hinder) engineers to do their work. Consequently, interest has increased regarding studies of and publications on scholarly communication and information exchange processes and systems since 1994. Many of these are directly applicable to engineers.

This book synthesizes the historical context surrounding early studies on the communication practices of engineers and scientists; looks at various aspects of communication through scientific and technical information (STI); examines the literature that distinguishes the information needs and uses of engineers from those of scientists; and offers a review of significant studies and projects that explore the communication practices of engineers.

The 1950s witnessed several excellent studies of how engineers and scientists communicate; however, research and surveys on the relationship between scientific activity and STI research took off in the 1960s, largely due to funding from the U.S. federal government and governments in Europe. The 1970s and 1980s saw a continuation of these studies, although this research slowed down by the early 1990s. Most of these studies defined communication broadly to include the creation of knowledge and its preparation for dissemination, the numerous channels by which it could be transmitted, and the assimilation and use of information the engineers received. Various meanings to the terms "information needs," "information seeking," and "information use" are found in the literature. For example, to some communication researchers "information needs" refer to the sources of information used, while for other researchers, "information needs" apply to the information content needed by engineers. Still others define "information needs" as the reasons for needing information.

Five types of models were used to examine STI communication in communication research since 1970. These models either:

1. Focus on communication during research and development projects and tasks; or
2. Follow the flow of information between individual engineers; or
3. Track information through its life-cycle; or
4. Examine the amount of information activity and use involved in specific work activities or by specific participants; or
5. Measure the amount and characteristics of information flow between various functions and participants.

It has been well documented over several decades that engineers spend much of their time communicating. This is often done to enhance their professional performance, as there is ample evidence of a correlation between engineers' communication and their work performance. However, the importance given to different types of information (e.g., literature versus interpersonal exchange) being communicated varies among studies. Furthermore, choices from among information sources are often dictated by factors, such as ease of use or cost considerations.

Many studies found that personal and interpersonal information sources are used initially by engineers and that internally published technical reports are favored over externally published documents. For this reason, uses of journal articles, books, and other sources of externally published material were given less emphasis by communication researchers. Later research began to focus on the importance of journal articles and discovered that engineers in universities read scholarly articles a great deal and engineers elsewhere read them less frequently, but value them nevertheless. Research and engineering education also began to focus on the importance of writing, presentation, and other communication skills.

The research on secondary sources of STI during this period was as extensive as that on primary sources. Most of the studies on secondary sources focused on automated bibliographic searching, with little attention on printed indexes or numeric databases. Studies from the 1960s dealt with the quality of output from information retrieval systems. Studies in the 1970s and 1980s of automated bibliographic databases tended to address evaluation or research involving system innovation and on "end-user" searching. Library resources and librarians were shown in the literature to be "under-used" by engineers in the completion of major projects. Libraries often fill a niche in the communication process, however, by providing for special needs, such as identifying and providing access to older or costly material.

There were many extensive reviews of engineering communication and related literature throughout this period. These include chapters in the *Annual Review of Information Science and Technology* (Menzel, 1966b; Herner and Herner; Paisley, 1968; Allen, 1967, 1969; Lipetz, 1970; Crane, 1971a; Lin and Garvey, 1972; Martyn, 1974; Crawford, 1978; Dervin and Nilan, 1986; Hewins, 1990; King and Tenopir, 2000); several books (Pinelli, Barclay, Kennedy, and Bishop, 1997a,b; Griffith, 1980; Kent, 1989; Nelson and Pollock, 1970; Mikhailov, Chernyi, and Giliarevskii, 1984; Williams and Gibson, 1990; Hills, 1980; Katz, 1988; Tenopir and King, 2000a), reports such as those produced by Pinelli and colleagues and King with Casto and Jones; and PhD dissertations such as Raitt.

Studies concerning STI communication often do not make the distinction between scientists and engineers. Authors who discussed the variations between the two groups before 1994 include Gould and Pearce (1991), Blade, Rosenbloom, and Wolek (1967), Allen (1988), and Pinelli (1991). Engineers were found to rely more on informal and interpersonal information sources than of published literature (Rosenbloom and Wolek 1967; Allen 1988) and they also read fewer journal articles and use the library less than scientists (Griffiths, et al.).

Several sustained and exemplary STI communication research projects were performed from the 1960s through the current time. All of these studies have relied heavily upon data collected from statistical surveys of engineers. The first of these studies, by William Garvey and colleagues at The Johns Hopkins University, began in the early 1960s and lasted until the 1970s. Their work had two major foci. First, they were interested in the "flow" of STI through various communication channels such as internal reports, professional meetings, journal articles, and so on. They developed a timeline to show when created information would appear in each of these channels. Second, they examined which sources of information engineers used for completing their work activities.

Thomas Allen and his colleagues at the Massachusetts Institute of Technology performed another series of studies initiated in the mid-1960s which continued into the early 1990s. Their work involved "record analysis" and self-administered questionnaires of engineers, which revealed that there are often individuals in an organization known as "stars" or "gatekeepers" upon whom others heavily rely on as sources for internal and external information. They identified nine basic information channels and determined the extent to which each of these channels are used, the value of these channels, and the factors which lead to their use.

King Research performed statistical descriptions of STI from the 1970s to the 1990s. Under National Science Foundation (NSF) contracts, King Research performed a series of studies to develop statistical indicators of STI. This research provided trends and projections for STI literature, libraries, authorship and information use by scientists and engineers, and STI expenditures in the United States. One finding debunked the myth of an "information explosion." Rather, growth in the literature merely reflected a growth in the number of scientists and engineers, a fact that holds true today. In 1976, they began research on the feasibility of electronic publishing of journal articles and concluded that the short-term future would have a two-tier system of dissemination (print and electronic). Results from the journal studies led to a book (King, McDonald, and Roderer, 1981) in which the entire journal system is described in detail. They then started research in 1981 to explore the use, usefulness, and value of STI and the contribution that STI services make to these outcomes. From the 1980s to the late 1990s, King Research performed numerous proprietary studies in various organizations to determine the communication activities of professionals (including scientists and engineers). Their work found that engineers and scientists spend a majority of their time communicating. They also found that engineers and scientists use a variety of information sources with choices being dictated by economics among other factors (new analyses from these studies and more recent comparative data are included in several chapters in this book). A continuation of these studies is being continued at the University of Tennessee (Tenopir under SLA, EIF, and other sponsorship), Drexel University, and University of Pittsburgh.

From 1977 to 1981, Hedvah Shuchman and colleagues of The

Futures Group conducted surveys of engineers employed at 89 firms. Sponsored by the NSF, these surveys examined the steps used in locating information needed to solve a project or task. The most important steps were personal stores of technical information, informal discussions with colleagues, and discussions with supervisors. They also found a discrepancy between the sources of information used and sources of information produced.

Beginning in the early 1980s and continuing into the 1990s, Thomas Pinelli, John Kennedy, Rebecca Barclay and their colleagues examined the diffusion of knowledge through the aerospace industry. Their work was undertaken as the *NASA/DOD Aerospace Knowledge Diffusion Research Project* and was done in collaboration with the NASA Langley Research Center, the Indiana University Center for Survey Research, and Rensselaer Polytechnic Institute. The project tracked the flow of STI at the individual, organizational, national, and international levels and examined the communication channels in which STI flows and the social system of knowledge diffusion. More information on the NASA/DOD Aerospace Knowledge Diffusion Research Project can be found in Chapter 12, which is dedicated entirely to this extensive research.

The data for this book are derived from many sources. A primary source is from readership surveys performed by King Research and the University of Tennessee School of Information Sciences, totaling results from over 15,000 scientists. Conducted since 1974, these surveys looked primarily at journal readership, although use of library and other information services was also considered. Data also came from the tracking of 715 scientific journals over a 40-year period and numerous cost studies of scientists' activities, library services, publishing, and other processes relevant to the journal system.

1.2   STRUCTURE OF THE BOOK

In Chapter 2, we describe a few of the many models that depict engineering communication. The principal models presented here attempt to illustrate the complexity of communication processes, which consist of many interpersonal or oral channels (e.g., informal and formal discussions, presentations, lectures, etc.) and written or recorded channels (e.g., letters and e-mail, electronic engineering handbooks and manuals, documentation of work, conference proceedings, articles, books, patents, etc.). Multiple channels exist because each serves specific information needs and requirements. Some information passes through a multitude of channels over time and a model is presented describing the "life" of information through these channels. Some channels, such as those found in the literature, involve many important system-like functions and the participants who perform these functions. These relationships and the life cycle of information through the journal channel form the basis for other communication models that are changing with new technologies.

Chapter 3 discusses the interrelationships among the engineering professions and work performed, resources used to perform engineering activities, and the output from those work activities. Information, of course, is an essential input resource to the work process, as well as a tangible output from the work process. We emphasize that receiving and using information requires substantial amounts of engineers' time, as well as, the use of information seeking tools such as technologies and library resources. The same is true in information outputs such as in preparing presentations and documents.

Chapter 4 deals with the engineering profession and how engineers go about their work. Examples are given for the amount of time engineers (in industry and government) spend in various work activities and the relative importance of information resources used by engineers to perform these activities. We also discuss engineers' general communication practices and how well they adapt to communication innovations. The fourth chapter also investigates how engineers assimilate new information into the work process, how they revise their work to take advantage of it and what the outcomes are of using the information. New information may also render old information obsolete, or indicate new, previously unimagined possibilities for the use of old information. We also examine how new technology might improve how engineers communicate in the workplace.

In Chapter 5, we first examine information seeking and use by engineers. There are three stages to this process: information seeking, information receiving, and reading/listening. Engineers, having decided that they have a need for information, must attempt to find information that best suits their need. Both of these processes form information seeking. When they have identified some information, in the form of an article or conference proceeding, for example, they must then attempt to acquire the information. There are many possible avenues through which users can acquire information: from asking a colleague, using a library resource, to logging a formal request for document delivery with a reprint service. The third stage, reading/listening, is the incorporation or assimilation stage. There are many levels of incorporation. Sometimes people skim through an article, only reading the principal statements and glancing at the figures; other times people, read very thoroughly; and most times people do a combination of the two at different times. Some of our research and data reveal evidence about the way engineers read relative to scientists in other fields. There has been less research on listening by engineers; however, since so much of the communication by engineers is oral, studies of listening and understanding are of particular relevance to the education and research communities. In this chapter, we also describe the extent to which channels are used and how much time is spent in information seeking and use. Chapter 6 pays particular attention to factors that affect engineers' communication channels, such as geographic or cultural differences among engineers; differences among branches of engineering; nature of the work performed; organizational policies; and personal characteristics such as gender, age, and so on.

Chapter 7 explores the facets of output and communicating information. The two major aspects in this chapter are writing and presentations. This is the communication stage, wherein engineers disseminate the results of their research or engineering output to their colleagues or to the public. We explore trends in how engineers communicate information in writing, verbally, in conferences or presentations, or in formal education settings, such as classrooms. We provide estimates of the amount of communication (e.g., presentations made, proposals written, etc.) and the time spent communicating. Chapter 8 discusses how education and training are changing in order to improve communication skills of engineers.

Because of the importance of engineering journals and the changes due to electronic publishing, Chapter 9 is devoted to the engineering journal channel. In this chapter, we examine the trends in authorship, reading, information seeking patterns, and publishing of engineering scholarly journals. We also present specific readership data: How articles are identified and where they are obtained. In particular, we present survey results for engineers' reading patterns before and after electronic journals became available.

Since electronic publishing is making a profound impact on information seeking and reading patterns, we devote all of Chapter 10 to survey evidence of these changes. This chapter examines current (2000 to 2003) observations of the use, usefulness, and value of journals; where engineers now obtain their articles; how they learn about the articles they read; the format read (print or electronic); issues concerning the age of articles read; and factors that affect choices from among journals read, sources used, and means of identifying articles.

Chapter 11 examines differences in engineers' communication patterns and also differentiates between engineers and scientists and medical professionals. Knowing precisely how expectations and communication styles differ between cultures can inform and potentially improve collaboration between engineers in different regions. It is equally useful to examine how information use patterns vary depending on the gender, age, level of education, or experience of engineers. Work roles assumed by the same individual over time and specialization in different fields of engineering also affect how engineers use information, because the types of goals and the procedures required to meet them differ substantially with different work roles or branches of engineering. Even stronger differences exist between engineering as very practical and applied, and science, which can be more theoretical and experimental. Chapter 12 elaborates on the extensive work performed by Pinelli and colleagues, which was discussed earlier. Finally, Chapter 13 summarizes the findings and provides conclusions about the communication patterns of engineers.

2

COMMUNICATION MODELS

2.1 INTRODUCTION

Innovation never happens in a vacuum; innovation requires communication. Just as work on the cutting edge of engineering and
science has become more technical and complex, so too has the
process of communicating. In becoming so, communication has
unfortunately also become more complex and cumbersome for
many of the engineers. Engineering is increasingly collaborative,
multidisciplinary, and global, but the goals of engineering projects are becoming progressively more refined and specialized.
Generally, the more narrow the discipline and the more specialized the information needs of its practitioners, the more difficult
it is to find good information easily. Engineers are rarely taught
advanced techniques of information retrieval, however, and are
typically not naturally gifted communicators, making it difficult
to fill their complex information needs (which can then impair
their ability to produce high-quality work).

In the quest to make all stages of research, development, design, and production as efficient and effective as possible, it is
important to posses a clear understanding of how engineers determine their information needs, fill them, use the information,
and share their own resulting information. By discovering these
patterns and systematizing that knowledge, communication can

Communication Patterns of Engineers. By Carol Tenopir and Donald W. King
ISBN 0-471-48492-X © 2004 Institute of Electrical and Electronics Engineers

be improved and communication at all stages of engineering work can be made more effective. This, in turn, increases the potential for high-quality progress in engineering endeavors. This chapter presents some of the major conceptual communication models and publishing endeavors that laid the groundwork for understanding those processes. The remaining chapters focus more specifically on current issues of how engineers communicate.

2.2   MODELS OF COMMUNICATION SYSTEMS

Many scholars have studied the systems of communication and the processes of information exchange, both in general and in specific subject domains. Several contain conceptual models that portray these complex patterns. The SCATT Report, by R.L. Ackoff, et al. (1976), describes an ideal Scientific Communication and Technology Transfer (SCATT) system that can be scaled to regional, national, or international levels of use. (See Figure 2.1) Ideally, the SCATT system facilitates the movement of scientific and technical information in multiple forms (audio and visual), multiple registers (formal and informal), multiple levels (primary—new information; secondary—about the new information; and tertiary—about the content of other messages), and multiple stages (production, dissemination, acquisition, and use). Each of these forms, registers, levels, and stages is an element of communication and so must be examined both individually and in interactions with the other elements in any useful exploration of the subject.

For example, an engineer develops a new technology, writes the patent application for it, and makes a video demonstrating how it works (stage 1). The patent is awarded and the videos are sent to other engineers around the world (stage 2) where they watch the video (stage 3). The other engineers take this new information (level 1) and talk about it (level 2) with their co-workers around the water cooler (informal register), and may also present it during a project meeting (formal register).

They begin to theorize about what they discussed at the water cooler or in the meeting (level 3) and might test elements of this invention against their own ideas for making it even better (stage 4). As the information moves from the initial pool of new information through each stage of communication, the new ideas that it creates may result in new information that can be fed back into the original pool of information, thus starting the cycle again. Many communication systems rely on a selection of these elements, but a system that integrates them all could be used to provide scientists and engineers with whatever information they need, whenever they need it, and in whatever form would be most useful to them.

Garvey and colleagues at The Johns Hopkins University (Garvey and Griffith, 1972) and others have described the variety of channels by which scientific and technical information content is communicated. Some are oral in nature (e.g., oral reports, informal discussions, meetings and conferences), while other channels involve information recorded in documents (e.g., informal progress reports, technical reports, journal articles, books, and patent documents). What is particularly useful for their model is

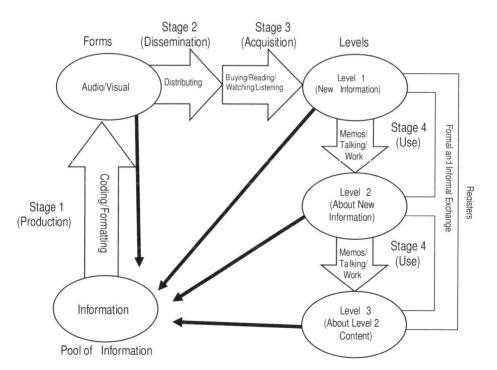

Figure 2.1   Scientific Information Transfer System Model. *Source:* Derived from SCATT Report.

that they observed time frames for the flow of specific information content through these channels from the time of creation to the time the information is reported by such means as laboratory notebooks, informal correspondence or interpersonal discussions, conference presentations and papers, research reports, dissertations, journal articles, patents, books, bibliographic entries, and state-of-the-art reviews. The schema in Figure 2.2 depicts a rough time frame of occurrence on the vertical axis (from the top down). However, less well documented in this context is the timeframe during which the information is obtained and applied by users as opposed to its first appearance in publications (Tenopir and King, 2000a).

Bear in mind the different aspects of communication and different channels of communication when exploring the interaction of information process cycles with communication systems. Scholarly journals are a useful example for demonstrating how these two phenomena interact, because they are a well-established method of formally communicating new information generated through research. Scholarly journals, like most well established products, became well-established because they work. Examining the development of something can give insight into the nature of the problem that it is designed to solve, so we offer a very short history of the scholarly journal. A more extensive history of print and electronic journals can be found in Tenopir and King (2000a) and Pullinger and Baldwin (2002).

2.3 MODELS OF SCHOLARLY JOURNALS

In the early stages of science, the scholars and practitioners were either dispersed over a wide geographic area or gathered in a few small areas. They usually communicated either face to face or in letters, which may or may not have circulated. As the number of scientists increased during the seventeenth century, scientific societies developed, designed to facilitate the exchange of information between members, even across national boundaries. These societies began to formally publish and distribute materials dedicated to their field. The scholarly journal soon became the centerpiece of the scientific society. As the numbers of scientists increased and the fields of study diversified, so did the scholarly journals. There was a marked increase in interdisciplinary science during the twentieth century because many of the sciences

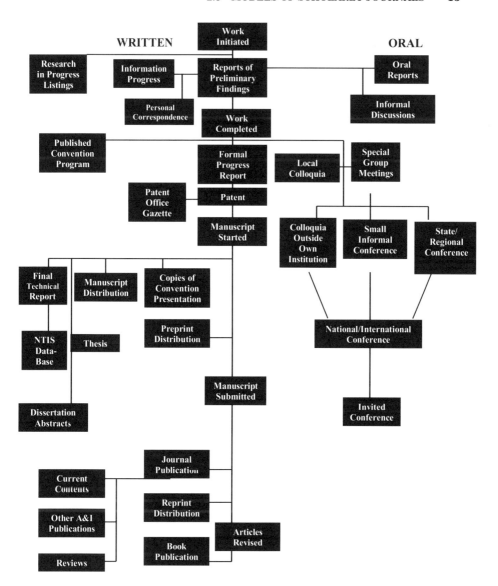

Figure 2.2 Communication Means. *Source:* Adapted from Garvey, Lin and Nelson (1970); Lin, Garvey, and Nelson (1970); and Garvey and Griffith (1972).

became more applied and were brought to bear on complex issues, for example, maintaining a viable natural environment while building habitats suitable for people. The elements that must be taken into account during an environmental impact study are diverse, lying outside a single field; therefore scientists and engineers from different fields had to collaborate. This trend shows no signs of slowing down, and has become increasingly global in nature. Collaboration remains the watchword of the twenty-first century and communication efficiency is as important today as it was in the seventeenth century.

The function of a scholarly journal is to package new, edited, and peer-reviewed information so that it can be transferred from one scientist to another, or to many. Several works can be collected and packaged according to type of content, which allows a journal to be tailored to the interests of its target audience. Packaging articles this way cuts the costs of delivery. Since the journals typically record both the works and the names of authors, they protect against plagiarism and enable recognition and prestige to authors and their institutions. The peer review facilitates trust and editing helps ensure quality. Because journals are widely distributed, it is less likely that information will be altered or lost. Authors who subject their work to the scrutiny of others are more likely to be conscientious in their research and disciplined in their writing, which leads to a higher standard of quality of information.

Scholarly journals have changed dramatically since their inception in the seventeenth century. The journal system became more complex, involving a number of specialized system functions and participants whose role in the system was to perform the functions. To depict these functions and participants, a spiral of the life cycle of information content in scholarly journals was developed as shown in Figure 2.3 (King, McDonald, and Roderer, 1981).

The spiral includes 11 functions, beginning with research and other sources of information creation (1). This function is the role of engineers and scientists. As a result of research, development, and other means of creating information, article manuscripts are composed (2) by engineers and scientists. The composition function refers to formal writing, editing, and reviewing of the manuscripts. When a manuscript is in a form to be communicated, it is recorded (3). These two functions are the role of authors, publishers, editors, and reviewers. At this stage authors have, as yet, very little impact on the scientific and engineering communities by means of formal communication. Only when the work has been reproduced and distributed does it gain the potential for widespread influence on an audience.

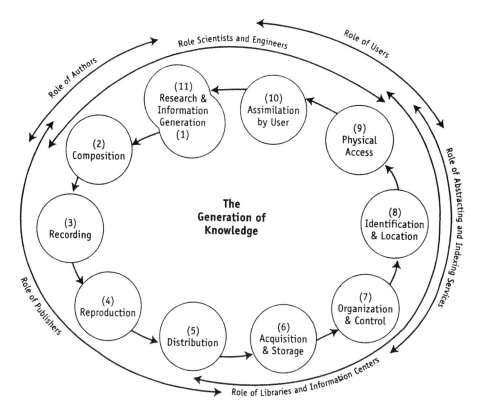

Figure 2.3   Life Cycle of Scientific Information Through the Scholarly Journal System. *Source:* King, McDonald, and Roderer (1981).

The reproduction (4) and distribution (5) functions are usually the role of the publishers; however, authors, libraries, and colleagues also play an important role. The transfer of documents through the three participants may be thought of as indirect reproduction and distribution, which requires acquisition and storage (6). Although many individuals acquire scientific and engineering articles and may store them, this stage of the spiral is represented by libraries and other information centers. Through their acquisitions and storage policies, libraries provide a permanent archive of scientific achievement. They also ensure access to this record.

Libraries play an important role in organizing and controlling these functions (7). In addition to collecting publications, libraries and other information centers provide access to these documents through classifying, indexing, and other related procedures. The major indexing and abstracting services and bibliographic services play an important part in organization and control as well. Needed publications may be identified and located (8) by a number of processes, including reference to one's own subscription, library search, and automated search and retrieval systems. This function is often accomplished for users by an intermediary from a library or other information service. The physical access (9) function includes direct distribution of reprints by the author. The function of assimilation by user (10) is the least tangible. It is the stage at which information content (as opposed to articles) is transferred. It is at this stage that the state of the user's knowledge is altered.

The functional schema is presented as a spiral because the communications process is continuous and regenerative. Readers may assimilate information they can use in their research in such phases as conceptualization, design, experimentation, and analysis. This research may, in turn, generate new composition (11) and recording for another cycle through the information transfer spiral. In a sense, the Garvey and Griffith model (Figure 2.2) describes the life cycle of information through the communications system of channels. This model, and the one in Figure 2.4, describes the life cycle of information through a particular channel—scholarly journals.

As electronic and digital processes migrated into the journal system, more functions became involved and more participants performed these functions. In 1994 the Association of American Universities Research Library Project expanded on the functions developed earlier for the traditional print journal. Furthermore, they identified three potential variations that they called classical, modernized, and emergent models, which differ in the type of format primarily used to transfer information. The variations on this model show the effects of these changes.

First, the AAU Task Force identified a system of scientific and scholarly communication that included: information generation and creation (data collection and analysis/synthesis), authoring (writing and revising), informal peer communication (informal editing and preprints), editing and validation (formal editing and peer review), ownership, privacy, and security (copyright issues), distribution (wholesale and reprints), acquisition and access (purchase), storage (holding), preservation and archiving (holding and conservation), information management (classification and biblio-

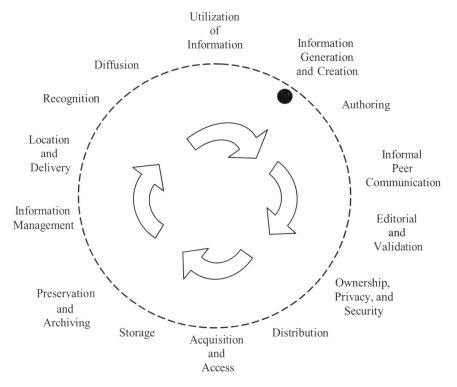

Figure 2.4 Scholarly Journal Information Cycle Model. *Source:* Derived from Association of American University Research Libraries Project Report, 1994.

graphic control), location and delivery (reference), recognition (author or institutional awards), diffusion (distribution outside the immediate community), and utilization of information by user. All models share similar elements, but differ in order of importance and ease of performance.

The *classical model* pertains mostly to print resources, such as print journals, printed conference proceedings, and so on. Print journals accommodate well the elements of authoring; editing and validation; ownership, privacy, and security; acquisition and access; storage; preservation and archiving; information management; recognition; and utilization. They retain author and institution information reliably and securely, can be acquired easily through established channels, stand up well to time and use, and their maintenance requirements are well understood. They are easy for most users to handle due to long familiarity with print formats.

The classical model is less well suited for handling the elements of informal peer communication, distribution, location and delivery, and diffusion. The print journals' performance in these areas suffers primarily because of time. It takes time to move a physical object from place to place, so print journals fare less well in these areas than electronic journals.

The *modernized model* also primarily describes print journal systems but concentrates on on-demand delivery of articles or journals and incorporates some use of electronic resources. The use of electronic versions speeds up the peer-to-peer communication and also the delivery and diffusion of the information. Since articles are usually then printed in hardcopy to read, they retain most of the classical assets of the print format. They lose some preservation and storage qualities because there is no control over the quality of paper used or the storage facilities available. Articles are the primary unit in electronic formats, so the benefits of having similar articles collected in journal form are not necessarily available.

The *emergent model* is based almost entirely on electronic means for communication and product formats. This has increased the information generated by increasing the ease and opportunity for collaboration. The electronic nature of the emergent model technologies raises issues of ownership, privacy, and security, because the intellectual property is easily copied and it can be difficult to control access and the integrity of the work. Preservation and archiving may be adversely affected, as well, because no one knows the most effective way to maintain electronic documents, so the information could be easily corrupted or lost. In many cases the storage of articles falls to the individual scientists or engineers, since physical copies may not exist in the library or information center. This may lead to a lack of organization for hard-copy versions of a work, which may impair access when the electronic documents are not available.

In the early 1990s, electronic journals began to be published in earnest, in CD-ROM and online. However, authors and readers were wary of their quality and sustainability and some raised the question as to whether journals were needed at all. Libraries were struggling with spiraling prices and pressures of physical space, but with a hope that the emergence of electronic journals might be the answer to these problems. Many publishers were hesitant to commit to electronic journals, but preprint archives developed at Los Alamos National Laboratory (LANL) gained widespread interest and, with the evolving technology and the emergence of Mosaic and the World Wide Web, seemed to trigger interest in electronic journals by all journal system participants.

By the late 1990s through the current time, authors and readers quickly accepted electronic journals as an alternative to print journals. In the 2003 online edition of Ulrich's International Periodicals Directory there are approximately 22,000 active, peer-reviewed titles, of which approximately 11,000 are available electronically. Most of these electronic journals are merely replicas of traditional print journals (some published exclusively in electronic format and most published in both formats). Of 797 scholarly engineering journals, all are available in electronic format.

During this time, libraries began to expand their collections of electronic journals in parallel with print or as a replacement to print. Some libraries have gone to nearly exclusive electronic collections. Most academic libraries began to rely on aggregator databases and/or negotiated licenses with publishers, library consortia, or other vendors. The LANL archives database moved to Cornell University (arXiv.org) and other preprint services also emerged. For example, the Department of Energy Preprint Network serves as a gateway to dozens of e-print servers (http://www.osti.gov/preprints/index.html). These e-print servers include preprints of articles submitted to peer-review journals, final versions of published articles (postprints), and articles never submitted to journals (Lawal). Separate electronic articles may also be accessed from an author's website or the authors' institutional repositories. Although still in the early stages of planning, university libraries are beginning to use the Open Archives Initiative (OIA) standard (http://www.openarchives.org) to build repositories of the intellectual capital of their faculty. One problem with this approach is that readers may want information organized by subject, not by geographical location of the author. The DOE preprint search service attempts to address this issue by means of a limited central search mechanism. However, with the form of the article and editorial standards controlled by the author, such preprint collections contain a variety of not necessarily compatible formats. Even within a complete journal model, there are many variations in e-journals. E-journals may be mere replicas of a print version, with papers presented in PDF format for easy printing but with poor searching capabilities.

Alternatively, they may provide a new e-design with added functionality, color graphics, motion files, and links to datasets. Browsing and searching may be offered or only one or the other. The availability of back issues also varies considerably (Tenopir, et al., 2003)

The processes by which the information in journals is made available and accessible to an audience are manifold. At the moment, for simplicity's sake, we will divide it into two parts: processes involving information content and processes involving information media. (See Tables 2.1 and 2.2.) To give a simple example, if an item is properly cataloged and classified, it probably can be easily found by the user. However, if that item is available only in a format that the user cannot access, then it is not useful to the person and the information might as well not even be there. It is important to coordinate these processes so that useful information is accessible to the right people when and how they need it.

2.4   MODELS OF INFORMATION SEEKING

Models of information processes, communication systems, and journal functions are not useful if the most important participant, the user, is omitted. One might well ask, "How do users know what, when, and how they need information?" Naturally, there

Table 2.1   Functions and Processes Involving Information Content

Information-Related Functions	Examples of Processes
Transformation	Translating from one language to another, subject or text editing.
Description and synthesis	Validating information through peer review, facilitating logical access through preparation of abstracts, indexes, catalogs, and metadata, preparation of reviews, especially state-of-the-art reviews.
Logical access	Identifying and locating sources through reference searching, referral, linking.
Evaluation/analysis	Assessment on behalf of users, annotated search output, data evaluation and integration by Information Analysis Centers.

Source: Tenopir and King (2000a).

Table 2.2 Functions and Processes Involving Information Media

Media-Related Functions	Examples of Processes
Communications	Information transfer, e.g., author to publisher, scientist to scientist, publishers to scientists, library to library.
Recording	Inputting to physical media, e.g., page masters, computer storage, CD-ROM disks.
Reproduction	Multiple copying, e.g., issues of journals, CD-ROM disks.
Physical transformation	Conversion, e.g., paper to microform, electronic to paper.
Storage	Providing access over time, e.g., libraries, computer files.
Preservation	Ensuring that information on the media or the media themselves do not deteriorate over time, and reproduction or restoration of information on deteriorating media.
Physical access	Delivery, e.g., personal subscriptions, library subscriptions, photocopies through interlibrary loan (ILL), terminal displays, computer printouts.

Source: Tenopir and King (2000a).

are models for information seeking patterns, as well. One of the most useful is Kuhlthau's "personal construct" model (Figure 2.5). She postulates that "information seeking is a process of construction that begins with uncertainty and anxiety. From a cognitive state of uncertainty concerning a problem springs emotional uncertainty." The distressed individual then relieves his or her uncertainty by finding and using information in the following six stages: (1) initiation; (2) selection; (3) exploration; (4) formulation; (5) collection; and (6) presentation.

The person decides that he or she needs something (1), focuses in on what the need is (2), explores what is involved in meeting this need (3), formulates a plan to meet all of the identified elements of the need (4), collects the information that will meet the need (5), and uses it (6). Note that this model shows a linear progression rather than a cyclical or iterative one. Each stage must be completed to the satisfaction of the individual before he or she moves on to the next. In the end, if all goes well, the individual is left with information that she or he can use.

Ellis (1982), following a different behavioral theory entitled the

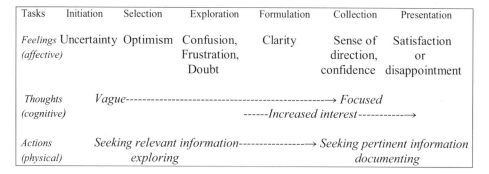

Tasks	Initiation	Selection	Exploration	Formulation	Collection	Presentation
Feelings *(affective)*	Uncertainty	Optimism	Confusion, Frustration, Doubt	Clarity	Sense of direction, confidence	Satisfaction or disappointment
Thoughts *(cognitive)*	*Vague--→ Focused* *------Increased interest------------→*					
Actions *(physical)*	*Seeking relevant information----------------→ Seeking pertinent information* *exploring documenting*					

Figure 2.5   Kuhlthau's Information-Seeking Model. *Source:* Kuhlthau (1993).

"grounded theory" approach, developed a model of information-seeking behavior for social scientists: (1) starting; (2) chaining; (3) browsing; (4) differentiating; (5) monitoring; and (6) extracting (Ellis and Haugan, 1997). (See Figure 2.6.)

Once again, the scientists in question were observed following a step-by-step approach to fulfilling their information needs. However, these scientists put in a browsing step, where they searched for information on general topics that might be related to their needs and then weeded out the irrelevant information in this pool of data. They were left with information relevant to their needs and could read it and extract the information that would serve them. Based on this research he went on to extend that model to the information-seeking behaviors of engineers and R&D scientists in the private sector in order to determine whether their processes are the same, similar, or different. The case studies will be discussed in later chapters.

Collaboration is increasing in today's scientific process and communication is a necessary part of collaboration. Therefore, it is worth investigating how scientists from different fields differ in their ways of communicating and how they differ in communication methods from engineers. By isolating differences and similarities the communication systems can have the flexibility to be useful to these different users. Gould and Pearce (1991) assessed the information needs of scientists compared to other kinds of scientists and also compared to engineers. They discovered that first and foremost engineers need quick access to current literature. Engineers are also willing to work in a more integrated informa-

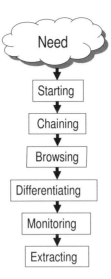

Figure 2.6 Ellis's Information-Seeking Model. *Source:* Derived from Ellis and Haugan (1997).

tion environment, using collections that contain full-text and graphics, for example. Therefore education of engineers should include instruction on the use of current and emerging information tools and resources.

Other studies, notably those by Allen (1960s–1980s), showed that engineers use their colleagues as a resource more often than scientists. Scientists use literature more than engineers. Also, engineers use the literature differently from scientists and often need different types of journals. For instance, engineers read more trade journals and internal reports than scientists do.

By collecting data on information-use patterns of scientists and engineers and comparing them to the information cycle, communication system, and user models that have been developed over the years, the strengths and weaknesses of the systems that are currently in place to serve the needs of engineers and scientists emerge. Identification of the weaker elements in these systems offers the possibility of modifying these processes, so, as a whole, communication becomes more efficient and effective.

3

A COMMUNICATIONS FRAMEWORK FOR ENGINEERS

3.1 INTRODUCTION

There have been numerous models of scientific and engineering communication over the past 50 years, only a few of which we described in Chapter 2. While useful in understanding communication processes, most of the models are conceptual in nature. Furthermore, they tend to ignore the consequences of communication for the work of scientists and engineers. Throughout this book we have tried to capture the essence of the conceptual communication models while providing a quantitative foundation for describing engineers' communication, as well as evidence of the consequences of communication processes for engineers' work and, in turn, higher-order effects derived from these processes. Of course engineers work varies depending partially on their work setting (e.g., industry, government, universities) and engineering discipline (e.g., electrical, aeronautical, civil).

As such, engineers can conduct research, design, develop products, construct, teach, manage, and perform other activities which require extensive resources. Such resources include engineers' time, support staff, computing and other equipment, instrumentation, and facilities, but critical, yet often overlooked, are information and information-seeking tools such as libraries and the World Wide Web. Not only is information an essential resource

Communication Patterns of Engineers. By Carol Tenopir and Donald W. King
ISBN 0-471-48492-X © 2004 Institute of Electrical and Electronics Engineers

for performing engineering activities, but the principal output from these activities is information in one form or another. This relationship is illustrated in Figure 3.1.

In this chapter we focus on the information inputs and outputs, and generally on how the information is communicated throughout the work processes. This framework depicts engineers' communication cycle as shown in Figure 3.2. At the heart of the engineers' communication cycle is the work performed by engineers, how information affects the work, and tangential relationships between information and engineers' work. Greater detail of these considerations is given in Chapter 4. However, at this point we define information input use (or receiving) by the effort or time engineers spend assimilating information through reading, listening, and so on, as well as the amount of reading or number of interpersonal contacts made. We describe this component of the communication cycle and factors affecting the use of information in Chapter 5.

Similarly, we define information output as the time and effort required for writing or making presentations and the amount of items written, presentations made, and so on. In Chapter 7 we

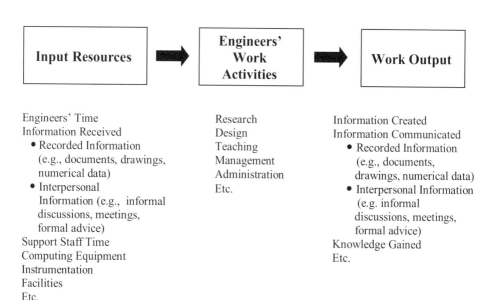

Input Resources	**Engineers' Work Activities**	**Work Output**

Engineers' Time
Information Received
  • Recorded Information
    (e.g., documents, drawings,
    numerical data)
  • Interpersonal
    Information (e.g., informal
    discussions, meetings,
    formal advice)
Support Staff Time
Computing Equipment
Instrumentation
Facilities
Etc.

Research
Design
Teaching
Management
Administration
Etc.

Information Created
Information Communicated
  • Recorded Information
    (e.g., documents,
    drawings, numerical data)
  • Interpersonal Information
    (e.g. informal
    discussions, meetings,
    formal advice)
Knowledge Gained
Etc.

Figure 3.1   Engineers' Work Activities, Input Resources, and Output.

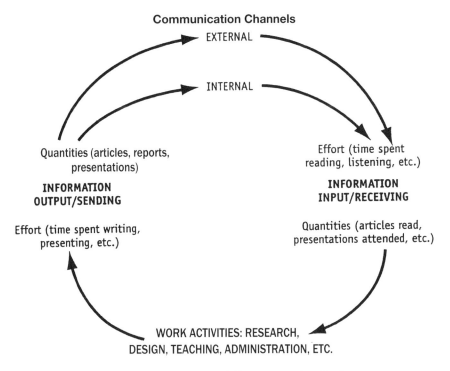

Figure 3.2 Engineers' Communication Cycle.

discuss the importance of engineers' writing and ability to express ideas and the essence of their work.

Generally, engineers appear to spend more of their time "outputting" information than "inputting" it. For example, Pinelli, et al. (1989) showed that engineers in aerospace spent 31% of 40 hours per week inputting information and 35% in output. The King Research surveys of engineers in industry and government showed they spent 9.7% of their time in informal discussions (receiving and communicating/sending), 21.1% in input/receiving, and 27.3% in information output/sending. How engineers communicate and the channels within which they do so are both important issues.

3.2   TIME ENGINEERS SPEND COMMUNICATING

Numerous studies corroborate the claim that engineers spend a majority of their time communicating (Hailey, 2000). Estimates usually range from 40 to 66% of their work time (Hertzum & Pejtersen), but may be as high as 75% (Nagle, 1998). A recent study of members of IEEE found that electrical engineers spent about 55% of their workday communicating. The majority of their communication was with other employees working on the same project (58%). Less time was spent communicating with other in-house employees (2%) and individuals outside the organization (22%). With in-house employees, most of engineers' communication time was with co-worker peers (62%) with less time communicating with supervisors (24%) and subordinates (14%). Forty percent of the communication time was spent in face-to-face communication, while the rest was spent writing, using e-mail and telephone, and interacting in small groups. Formal presentations account for only 4% of the communication time. Over half of external communication was by telephone (Vest, Long, and Anderson, 1996). One study suggested that engineers in industry may spend as much as half of their time solely on the task of writing (Dyke and Wojahn).

Estimates of the amount of time spent by engineers communicating are not new. For example, in 1984, Raitt surveyed European aerospace scientists and engineers and observed indicators of the amount of their time spent communicating. He reported that "over the six-month period in question 50% of the [survey] respondents spent less than one-third of their time communicating, while 8% spent over two-thirds." In 1989, Pinelli, et al. reported that aerospace engineers spent about 66% of their time communicating, although later (1991) they reported a smaller proportion of time (about 50%). Over a 12-year period (1986–1998) King Research surveys involving engineers in industry and government yielded estimates of 58% of their time spent communicating.* These surveys estimated that engineers spent about 2,130 hours performing work-related activities (not including vacation, holidays, and sick leave). The surveys also provided evidence that engineers may currently be spending more of their time working than they did earlier and most of the additional time appears to be due to an increase in communica-

---

*Ten studies of scientists by others (King and Tenopir) from 1958 to 1998 provided estimates ranging from 25% to 67% of time spent communicating (with an average of 47%). The King Research surveys ($n = 252$) asked respondents to indicate how much time (over a 40-hour week) they spend in work-related activities (2,128 annual hours total, not including vacation, sick leave, etc.) and what proportion of this time is spent in various activities.

tion activities. In support of this assertion, Pinelli, et al. (1991) asked engineers in several countries whether there were changes in the past five years in amount of time spent communicating technical problems. In the United States, 42% said time increased, 45% said it stayed the same, and 13% said it decreased. In Japan results were similar, but in Western Europe, 60 percent said the time increased.

Another issue is the growth in the amount of information, both from external sources as well as information found in engineers' organizations. The result of more information is that engineers must spend more of their time seeking relevant information. Although the studies mentioned earlier show that engineers spend up to 75% of their time dealing in some way with information, Court, Culley and McMahon (1997) estimate that they spend 20 to 30% of their time just searching for information. Others, such as Rzevski, argue that the 1997 estimate is too low; in fact they estimate much more time is spent in information seeking, maybe as much as 70% of an engineer's time. Much of this time is not used efficiently, however, and since engineers typically need information within short time frames, delays result in uninformed decisions. Most agree that the amount of information needed and the amount of time spent seeking information is increasing.

3.3 ENGINEERS' COMMUNICATION CHANNELS

There have been a number of characterizations of engineers' communication channels, several of which are presented below. In the mid-1960s, Thomas J. Allen began a series of studies which continued into the 1990s at the Massachusetts Institute of Technology (MIT) under various grants, some of which were from the National Science Foundation (NSF). Allen and his colleagues observed exchanges or flows of information between individuals, who were referred to as "stars" or "gatekeepers," whom others depended heavily upon for internal, as well as external, sources of information. These stars were particularly familiar not only with internal technical reports and the published literature, but also internal and external interpersonal sources of information. Allen also made a point of distinguishing information-seeking behavior of scientists from engineers. As a result of these studies, Allen and colleagues made suggestions as to how R&D organizations should be structured in order to optimize communication processes, particularly involving engineers.

Allen and colleagues identified nine basic information channels:

- *Literature:* Books, professional, technical, and trade journals, and other publicly accessible written material.

- *Vendors:* Representatives and/or collateral materials of suppliers or potential suppliers of design components.

- *Customer:* Representatives and/or collateral materials from the government agency for which the project is performed.

- *External sources:* Sources outside the laboratory or organization which do not fall into any of the above three categories. These include paid and unpaid consultants and representatives of government agencies other than the customer agency.

- *Technical staff:* Engineers and scientists in the laboratory who are not assigned directly to the project under consideration.

- *Company research:* Any other project performed previously or simultaneously in the laboratory or organization regardless of its source of funding. This includes any unpublished documentation not publicly available, and summarizing past research and development activities.

- *Group discussion:* Ideas that are formulated as the result of discussion among the immediate project group.

- *Experimentation:* Ideas resulting from test or experiment or mathematical simulation with no immediate input of information from any other source.

- *Other division:* Information obtained from another division of the same company.

In-depth research by Allen and colleagues determined the use of these channels, which ones appear to be most useful, and identified the factors leading to their use.

In 1967, Rosenbloom and Wolek surveyed more than 3,000 engineers and scientists in large corporations and from a sample of members of the Institute of Electrical and Electronics Engineers (IEEE). One principal focus of the study was to determine which information sources engineers use. They chose to categorize these information sources as to whether they were internal or external to the engineers' organization and whether the source was written (e.g., reports, professional or trade journals) or interpersonal (i.e., oral). Pinelli and colleagues (1990a) use a similar breakdown of communication channels and provide details of the documented channels with some emphasis on government technical reports. King Research surveys (1986–1998) similarly categorized channels as to whether they were internal or external and whether they were written (e.g., reports, articles, proposals, programs, e-mail, etc.) or interpersonal (e.g., informal discussions, consultations, meetings, and conferences, etc.). Garvey and Griffith present a similar breakdown (see Chapter 2).

Some scholars have categorized channels by the written and interpersonal breakdown, but also include tools used in information seeking such as libraries, the Internet, online information retrieval, and current awareness. For example, Shuchman surveyed 1,315 engineers (1977–1981) to examine steps taken by engineers in looking for information thought to be needed to work out a solution for the most important technical project or task being worked on. Such steps included personal stores of technical information, informal discussions with colleagues and supervisors, library sources, databases, and various technologies and media. Such channels were chosen for both input and output of work. In 1991, Gould and Pearce categorized engineers' channels as the primary literature (e.g., serials, patents, technical reports, standards), indexing and abstracting services (print and electronic), current-awareness services (e.g., current research, electronic networks, conference proceedings, letters, journals, newsletters, technical reports, preprints, databases), and other electronic sources. More recently, Harris discusses communication in a networked-based systems engineering environment.

Allen and his colleagues have devoted a great deal of research to examining flows of information within organizations, particularly how some individuals serve as conduits to information. Their research methods have been replicated by a number of communication researchers. For example, Tushman undertook a series of studies examining communication in organizations as part of his doctoral thesis under Allen's supervision. His most extensive research involved a survey of professionals in an R&D facility of a large corporation. In this survey he relied on a "personal contacts record" for 15 weeks, in which data were recorded one day a week on specified days. More than 400 professionals were surveyed in total. Tushman studied several dimensions of communication: the type of work being performed (i.e., basic and applied research, development, and technical service), level of dependence on information (intraproject, intrafirm, and extrafirm), environment in which task is performed (i.e., stable or turbulent), and perception (by others) of the projects as being high- or low-performing. He also examined the relevance of (1) information "stars," who are approached as an information source with high frequency by colleagues, (2) "boundary spanners," who span communication boundaries between units in an organization or between projects in the organization and the outside, and (3) "gatekeepers," who are both information stars and boundary spanners.

One of the most detailed descriptions of communication channels used by European aerospace engineers was presented by Raitt (1984) as part of his Ph.D. dissertation. An adaptation of Raitt's way of characterizing engineers' communication channels is given in Figure 3.3. This schema, while relatively straightforward, begins to illustrate the complexity of engineering commu-

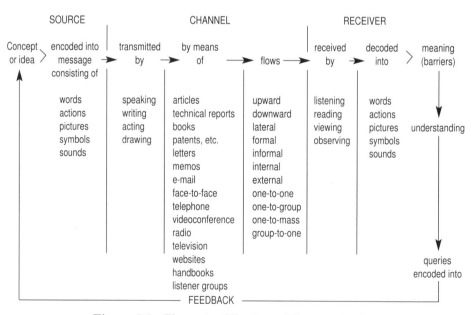

Figure 3.3 Elements of Engineers' Communication

nication. It does lack the element of the purposes for which information is communicated, and therefore is just an extension of the upper half of the communication cycle given in Figure 3.2.

In one type of specialized communication, preparing patents, engineers have a distinct and identifiable information use pattern. Breitzman (2003) found that information technology patents increasingly reference scholarly papers, in particular publications of the IEEE. In turn, patents that cite IEEE publications are cited more frequently than other patents.

3.4   FACTORS AFFECTING ENGINEERS' COMMUNICATION CHOICES

Richard Orr observed that people can get needed information (1) by experimentation, observations, and so on; (2) from contemporaries' brains; and/or (3) from the pool of recorded knowledge (i.e., the past effort of others). The choice among these three options depends on the perceived likelihood of success within an acceptable time period and on perception of relative accessibility, cost (i.e., time and expenditure), and effort necessary to obtain the information. Choices are also made by weighing the amount of information versus cost (Griffiths and King, 1991; Tenopir and King, 2000a) and by considering what is "good enough" or "not good enough" versus some variables that contribute to this assessment (Orr, 1970). On the other hand, Allen and Gerstberger (1964) found that engineers appear to act in a manner not primarily to maximize gain (i.e., benefits) but rather to minimize loss (i.e., cost). Yet, Meadows summarized the many surveys done over the years as follows: "One of the firmest conclusions of information usage surveys seems to be, indeed, that the intrinsic value of an information channel has little, or no bearing on the frequency with which it is used. The ultimate factor is always its accessibility."

Orr suggested two types of variables or factors related to choice of communication channels.

1. *Personal factors:*
   - Education, training and past work, including discipline/profession, level of training, nature of work, experience with channels

- Status and stage of career
- Demographics
- Inherent capabilities
- Personality/work style

2. *Situational factors:*

- Nature of need, including functions served; kind of information (i.e., theories, methods, data/results); information attributes (i.e., precision, quality, specificity, complexity, urgency)
- Current project, including nature of work, stage of the project
- Work setting, including structure, reward/control system, size of organization, information infrastructure, prestige of setting
- Sponsor/funder characteristics
- Peer communities
- Channel capabilities and attributes

We discuss these factors further in Chapters 5, 6, and 7.

4

THE ENGINEERING PROFESSION AND COMMUNICATION

4.1 THE ENGINEERING PROFESSION

Engineering has been called an applied science that has as its goal the creation or improvement of products or technologies with immediate practical application. Unlike scientists, engineers typically work on projects assigned by their management, most often work in industry or government, conduct projects in teams, and focus their goals on company or organizational success (Von Seggern and Jourdain, 1996).

Engineers tend to be specialists who work to solve technical problems collaboratively with other engineers and with scientists. Typically engineers in both industry and academia "need more information than they generate," rely on personal information and experience, seldom use libraries, and use more texts, technical reports, catalogs, and trade journals rather than scholarly journals or conference papers (Leckie, Pettigrew, and Sylvain, 1996).

The engineering profession is unique in many ways. For example, engineering is "context-specific and often involves proprietary information" and, as a result, "engineers tend to rely on conversations with internal colleagues and clients" (Veshosky, 1998). Scientists, on the other hand, are accustomed to external communication sources, including academic journals and free exchange

Communication Patterns of Engineers. By Carol Tenopir and Donald W. King
ISBN 0-471-48492-X © 2004 Institute of Electrical and Electronics Engineers

with colleagues at other organizations. Engineers also communicate some with external colleagues, but to a much more limited extent. The thinking process required in engineering requires more complex and abstract problem solving than most disciplines, so communicating with non-engineers may be difficult (Li, 1994).

Although projects are assigned, engineers typically have freedom in deciding how to do their work and "they are expected to make informed decisions in a number of situations where many possible solutions are available" (Hertzum and Pejtersen, 2000). The choices that engineers make in deciding how to approach their tasks and solve problems depends on their "understanding of the context of the task and, consequently, on their success in obtaining information about this context" (Hertzum and Pejtersen, 2000).

4.2   HOW INDUSTRY AND GOVERNMENT ENGINEERS SPEND THEIR TIME

The King Research surveys of engineers (1986–1998) in industry and government revealed that engineers spend about 2,128 hours in work-related activities (as shown in Table 4.1).

Most of the time spent by these engineers is in direct engineering activities such as design and drawing (28%), primary research such as data collection, observation, experimentation, etc. (13%), and technical activities such as research support or secondary or background research (15%). They also spend some time in their own professional development (6%) or educating others (6%). However, a substantial amount of time is spent on such activities as management (16%), budgeting (2%), marketing (2%), and the like. It may be that formal education of engineers should acknowledge or prepare engineers for these kinds of activities.

4.3   IMPORTANCE OF INFORMATION RESOURCES TO ENGINEERS' WORK

The studies by King Research above (1986–1998) examined the importance of information and communication tools to the work activities (given in Table 4.1). An indicator of information importance was derived from questions asking engineers to rate the im-

Table 4.1 Average Annual Amount (Hours) and Percentage (%) of Time Spent by Engineers in Performing Various Activities, 1986–1998

Type of Activity	Engineers	
	Hours	(%)
Primary research (data collection, observation, experimentations, etc.)	281	13.2
Engineering (design, drawings, etc.)	593	27.9
Technical or research support	177	8.3
Secondary or background research	142	6.7
Management, supervision, hiring	336	15.8
Finance, accounting, budgeting, etc.	37	1.7
Operations, practice	83	3.9
Marketing & sales	41	1.9
Professional development	135	6.3
Educating & training others	136	6.4
Other	167	7.8
Total	2,128	99.9

Source: Surveys at AT&T Bell Labs, Air Products & Chemicals, Inc., Baxter Healthcare, Eastman Chemical Co., Eastman Kodak Co., National Institutes of Health, Procter & Gamble Co. ($n = 252$).

portance of various resources in performing these work activities. Ratings were from 1 to 5 (1 being "not at all important," 3 being "neutral," and 5 being "absolutely essential"). These results are displayed in Tables 4.2 and 4.3. Engineers used a number of different information-related resources to do their work, including computing equipment or PCs, other equipment or instrumentation, information found in documents (e.g., articles, books, patents, technical reports, etc.), support staff (e.g., secretaries, technicians, etc.), information staff (e.g., librarians, information specialists, etc.), and advice from consultants or colleagues.

Information found in documents rated as having more importance than other types of resources on three different types of work activities: primary research, secondary research, and professional development. In addition, information found in documents rated second in importance for engineering (design, drawings, etc.). Management activities, marketing and sales, and education activities were the only areas in which document information was not rated first or second. Computing equipment/personal computers tended to be rated highest in importance for engineering (de-

Table 4.2 Average Ratings* of Importance of Resources Used by Engineers in Performing Various Work Activities by Type of Resource: Observed 1986–1998

Type of Work Activity	Annual Time Spent on Activity (Hours)	Type of Resource					
		Computing Equipment/PCs	Other Equipment/Instrumentation	Information Found in Documents	Support Staff	Information Staff	Advice from Consultants/Colleagues
Primary research (data collection, observation, experimentations, etc.)	281	4.23	4.19	4.31	3.60	3.16	3.79
Engineering (design, drawings, etc.)	770	4.13	3.41	3.88	3.29	2.61	3.84
Secondary or background research	142	2.88	2.44	4.41	3.10	3.43	3.44
Management, supervision, hiring	373	3.02	1.69	2.79	3.07	2.35	2.90
Marketing & sales	41	3.13	3.00	2.56	3.00	2.14	3.07
Professional development	135	3.26	2.40	3.91	2.41	2.71	3.46
Educating & training others	74	3.50	3.53	3.38	2.67	2.71	2.81
Other	312	3.27	3.78	3.69	3.56	2.85	3.68

*Importance ratings: 1—Not at All Important; 3—Neutral; 5—Absolutely Essential.
Source: Surveys at AT&T Bell Labs, Air Products & Chemicals, Inc., Baxter Healthcare, Eastman Chemical Co., Eastman Kodak Co., National Institutes of Health, Procter & Gamble Co. ($n = 252$).

sign, drawings, etc.) and marketing and sales. The relative ratings among resources appear to be what one might expect, but the ratings do show that the importance of these resources is quite high for many activities.

In Table 4.3, the ratings of importance of different resources are given for different communication activities. Information found in documents rates highest for consulting and advising of others, and second highest for all other types of communication activity. For every activity, documents were rated as better than neutral (numbers averaging higher than 3). Computing equipment and advice from consultants and colleagues also tended to be relatively high in importance for communications activities. The average annual amounts of time spent by engineers on the various work activities are also displayed in the tables, giving an indication of the relative importance of the activities themselves.

Reading was found to be important for a number of reasons (King and Tenopir, 2000a). For example, Scott (1962, 1969) reported that literature served as the primary source of creative stimulation. Chakrabarti and Rubenstein (1976) found that the quality of the information as perceived by the recipient is a major factor in the adoption of innovation. Ettlie (1976) also found that the literature was the single most important source of information in achieving product innovations. Another aspect of the usefulness of scholarly journals is their importance to scientists. Machlup and Leeson (1978) report that economists found 32% of their readings to be useful or interesting, 56% moderately useful, and 12% not useful. University scientists rated importance of reading to their work (from not at all important—1 to somewhat important—4 to absolutely essential—7). Scientists and engineers rated the importance of the information to achieving teaching objectives as 4.83 on average, while importance to research was given an average rating 5.02. Over a period of a year, the scientists indicated that, of a total of 188 readings, an average of 13 readings per person were absolutely essential to their teaching and 23 were absolutely essential to research.

4.4 THE VALUE OF READING

Machlup (1980) points out that there are two types of value of information: purchase value and use value. The purchase value is

Table 4.3 Average Ratings* of Importance of Resources Used by Engineers in Performing Communication-Related Activities by Type of Resource: Observed 1986–1998

Type of Work Communication	Annual Time Spent on Activity (Hours)	Type of Resource					
		Computing Equipment/ PCs	Other Equipment/ Instrumentation	Information Found in Documents	Support Staff	Information Staff	Advice from Consultants/ Colleagues
Consulting/Advising others	222	3.27	2.73	3.69	2.84	2.50	3.57
Writing							
• proposals, plans	92	4.06	2.79	3.73	3.40	2.94	3.26
• technical reports	117	3.69	2.33	3.62	3.39	2.83	3.12
• articles, books, etc.	12	4.19	2.33	4.00	3.20	3.00	3.14
Making presentations	123	3.93	3.20	3.31	3.28	2.47	3.24

*Importance ratings: 1—Not at All Important; 3—Neutral; 5—Absolutely Essential.
Source: Surveys at AT&T Bell Labs, Air Products & Chemicals, Inc., Baxter Healthcare, Eastman Chemical Co., Eastman Kodak Co., National Institutes of Health, Procter & Gamble Co. (*n* = 252).

what users are willing to pay for the information in terms of money exchanged and the time expended in obtaining and reading the information, whereas use value is the consequence of using the information. The average purchase value expended per user on journals is at least \$6,000 per year and the use value exceeds \$25,000 per year per scientist or engineer (Tenopir and King, 2000a). The price paid in their time tends to be 5 to 10 times the price paid in purchasing journals and separate copies of articles.

There are many ways in which the consequences of reading can be expressed beyond dollars. For example, Tenopir and King (2000a) showed that nearly all reading by university scientists and engineers (95%) resulted in some favorable outcomes. Readings improved quality of teaching, research, or other activities for which the reading was done (66% of readings); and helped them to perform the activity better (33%), faster (14%), or at a lower cost in time or money (19%). With nonuniversity scientists and engineers, consequences were established for the principal activity for which the reading was done (Griffiths and King, 1993). For example, 67% of the readings resulted in higher quality, 32% in faster performance, 42% helped reinforce hypotheses or increased confidence in one's work, and 26% resulted in initiating ideas of broadened options concerning work.

There has been ample evidence over the years that there is a positive correlation between amount of reading and productivity of scientists and engineers. In the late 1950s Menzel related the numbers of journals read with productivity measured in number of publications. Orr cites several studies in the 1960s that suggest similar relationships. The Operations Research Group of the Case Institute of Technology concluded that publishing physicists and chemists read more than nonpublishers. Shilling, et al. (1964) established a strong positive correlation between various measures of the productivity of biological R&D labs and indices of communication. Allen (as cited in Nelson and Pollack, 1970) compared differences among engineering teams with regard to (1) the proportion of time spent with various types of input channels (and also the phasing of the use of these channels) and (2) the quality of their output of performance and of publications. Parker, et al. (1967) found that the strongest single prediction of production was the number of informal contacts with other scientists. Meadows (1974) cites Menzel, et al. (1960) as providing evidence that chemists who were rated as highly creative typically consulted twice as much literature as those having low creativity. Later Weil (1980) states that journals, when compared to other published materials and computerized information, provided by far the greatest benefits to current work. Griffiths and King (1993) established in each of six organizations that amount of reading is positively correlated with five indicators of productivity (i.e., outputs measured in five ways).

Another indicator of the use value of scholarly journals is that scientists whose work has been formally recognized tend to read more than others. For example, Lufkin and Miller (1966) in the 1960s report from surveys of engineers in two companies found that "People who have been singled out for excellence, whether by promotion, or by publication, or by special recognition for creativity, all read a great deal more than the average." Surveys by Griffiths and King (1993) in the 1980s and 1990s invariably showed that winners of achievement, technical, and patent awards read 53% more articles than non-awardees. Similar results were observed for those chosen to serve on high-level projects or problem-solving teams. In one company, 25 persons who were considered particularly high achievers read 59% more articles than their colleagues. Tenopir and King (2000a) report University of Tennessee scientists and engineers who received achievement awards or special honors read more; that is, those recognized for their teaching read 26% more articles and those recognized for research read 33% more articles. A similar result was observed in a survey of Drexel University faculty (King and Montgomery, 2002).

4.5   THE IMPACT OF INFORMATION ON INNOVATION

Engineers tend to be resistant to change, including adapting to new technologies and innovations. This characteristic reluctance to change extends to how they seek for or use information. Veshosky pointed out that the engineering and construction industry in the United States is perceived as being slow to innovate, unlike manufacturing industries that readily adopt computer-based innovations in design and management. The tasks that design engineers do on a daily basis make them some of the heaviest users of information and innovative information technologies, but construction site activities, on the other hand, "have seen relatively marginal changes in recent decades" (Veshosky, 1998).

Some possible reasons for this "resistance to innovation" are discussed by Veshosky and include:

- Concern about safety, quality, and costs
- Restrictive codes and standards
- The project orientation of the construction industry and the unique nature of individual projects, which complicates application of innovations across projects
- A fragmented industry
- The nature of the construction-related research and development systems, where most R&D is performed by universities, material or equipment suppliers, or governmental agencies, with little performed by industry firms

The construction industry in Holland in the 1990s was a mature and important industry poised for change. Pries and Janszen (1995) found a low level of innovations from within the industry. Instead, external forces such as deregulation, tensions between specialization and diversification, and price competition drove most innovations. According to Pries and Janszen, innovations are dependent on changes in management and a move toward a more extroverted and market-oriented approach. They believe education will help make these changes possible, although access to high-quality external information and effective use of it will certainly also be needed.

Baldwin and Sabourin studied Canadian food processing companies to compare those that innovate with those that do not. The authors found that there can be a "production-based" approach to innovation, not just an R&D approach. However, firms engaged in R&D are much more likely "to innovate" and an R&D "performer" has over an 80% chance of being an innovator, while a non-R&D "performer" has only a 50% chance. Innovations can be made in products, processes, or both, but R&D is important for product-only innovations. The success of innovation is often measured by increased profits, market share, and so on (Baldwin and Sabourin, 2000). Although they did not explicitly measure information use, it may be implied that the engineering focus that needs and uses more information (R&D focus) is more likely to initiate innovations.

By the 1990s, the manufacturing industry began to change due to pressures of increasing demand for higher-quality products, shorter lead-times, and the need for more flexible collaboration (Carstensen, 1997). In response to these pressures some manufacturing firms adopted "concurrent engineering," which introduces knowledge from all stages of the product life cycle into the earliest stages of a project. This means that many people with disparate competencies such as design, development, production, and use work together in teams to make design decisions, incorporating as broad a perspective as possible (Carstensen, 1997).

Carstensen studied a Danish manufacturing firm to discover what information engineering designers used in their design decisions and how they acquired access to that information. Figure 4.1 illustrates the most important interaction between the design team and the various stakeholders in the project. Designers sought different types of information from different sources at different stages of the project. Carstensen found that the designers used both internal and external information, personal and formal communication channels, and written and oral information. To make these sources most useful in the work process they needed to meet certain general requirements, which included:

- High degree of flexibility in mixing the use of many tools
- Access across different media
- Support of converters between different domain languages to bridge vocabulary problems
- Refinement of search profiles
- Search support

4.6 ORGANIZING INFORMATION FOR BETTER USE IN THE WORKPLACE

Information is considered critical for success in manufacturing and "is essential to economic vitality and growth." It is particularly important to organize and manage information in a way that accommodates the information-seeking needs of industrial engineers (Mathieu, 1995). Mathieu recommends the integration of information technologies within an infrastructure of communication networks, hardware and software applications, databases,

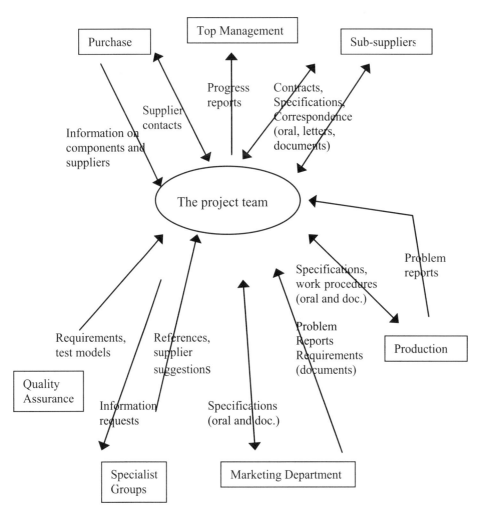

Figure 4.1 Context Diagram Between Project Team and Stakeholders. *Source:* Cartensen (1997).

bulletin boards, standard reference data on materials used in manufacturing, numerical data, and other information services.

   To be incorporated into their work process, information sources must be organized to match the way engineers do their jobs. For example, design engineering involves special information needs that require different types of information seeking than, say, construction engineers. Tasks that require high degrees of problem solving require a different style of information seeking "than does keeping up-to-date about potentially relevant innovations. Information seeking for keeping up-to-date is more general than for problem solving and should provide a basis for selecting among available sources when searching for information to solve a specific problem" (Veshosky, 1998).

Effective communication of and access to scientific and technical information has been shown to "play a critical role in the innovation process" (Von Seggern and Jourdain, 1996). Design and other similar engineering tasks rely on teamwork and collaboration at all stages. Systems that encourage better communication among team members will promote the efficiency of the design tasks and impact the success of such projects.

Design and R&D engineers need efficient access to information and communicate more than the average engineer. Because they are often constrained by time, work in teams, and work for a client, design and R&D engineers need up-to-date, accurate, and original information. According to Leckie, Pettigrew, and Sylvain (1996), this need for up-to-date and original information "explains why the journal literature is considered largely irrelevant" by design and R&D engineers. Instead they rely on internal technical information such as benchmark test results. The information needs in R&D vary with the stage of the project, but R&D engineers rarely search online bibliographic databases. Leckie, Pettigrew, and Sylvain suggest this may be because databases do not match engineers' way of working. Kahin suggests that organizations that rely on cooperation should collect and disseminate research communication.

Consulting engineers also consume more information than average, and in fact "are among the biggest consumers of information in engineering." Consultants also need current and accurate information, but they need external market information about vendors and customers (Leckie, Pettigrew, and Sylvain, 1996).

Team decisions within the various project stages should be recorded and stored in a way that makes the information easily accessible to others throughout the project life cycle and throughout future iterations, thereby effectively integrating the communication process into the design process. Well-documented and accessible decisions will save time and improve future projects. A study of 200 engineering designers in the United Kingdom found that design decisions were recorded in many places, including diaries (7%), memos (19%), reports (30%), logbooks (19%), data/calculation sheets (17%), drawings, and computer files (although 8% did not record their decisions) (Court, Culley, and McMahon, 1994b). A vast majority (86%) had access to a personal computer, but they preferred using paper both to record their decisions and to transfer the design intent to manufacturing.

It seems reasonable that integration of information resources would likely improve how engineers use these resources. Engineers need a variety of information sources, must access information quickly, and need different information at different stages of projects. Coordination of various formats (paper, computer-based, and human communication) is desirable (Baird, Moore, and Jagodzinski, 2000). The structure of work into orderly teams helped software engineers organize to find solutions to their project (Button and Sharrock, 1996). Organizing information resources for engineers in ways that mimic this natural structure may make them more useful for the engineers and, hence, more often used in their projects.

Although he does not specifically focus on engineers, Hertzum (1999) described how personal documents created internally were used to further the work of professionals in companies. Because many engineers work in teams, studying their use of personal documents can offer insights into how to design better computer-based document management systems that will facilitate sharing of the information stored in documents created during personal and group decision-making.

Hertzum identifies six roles that such documents play in facilitating a professional's work. Personally created documents in the workplace serve:

- As personal work files (that must be readily available to the individual and to colleagues within the organization)
- As reminders of things to do
- To share information with some, but to withhold it from others
- To convey meaning
- To generate new meaning
- To mediate contacts among people

Such documents help to record personal decisions and bring informal communication into the work process, but they are rarely integrated into an organization's information systems. As with oral information, the content of written informal documents is therefore lost if a person leaves his or her job or if personal memory fails unless a system that facilitates archiving and sharing of this important information is instituted. Clearly such a system can help in today's collaborative engineering environment.

Personal files and personally created information are essential to the work process. Professionals interact with their own information on three levels. Hertzum describes these levels as: *action information* (which includes documents readily at hand, often on a person's desk); *personal work files* (within reach, but usually on shelves or in filing cabinets); and *archive storage* (information stored away from the office).

Surprisingly, although most of such documents are now created on a computer, the computer is rarely used to store or retrieve them. Hertzum suggests this may be because people prefer to store their own documents together with external documents, such as correspondence or other materials that pertain to a specific project. Paper copies of their personal documents are printed out for storage and retrieval. Whether electronic or print, professionals tend to store their documents in ways that match their work projects, either by location or by the dictates of their task at hand (Barreau, 1995; Barreau and Nardi, 1995; Hertzum, 1999). An attempt has been made in Japan to provide a portal site that can be used by engineers to store information in a database and create discussion forums (Yamada, 2001).

Understanding how engineers choose and value information also will help design better information systems. Information-related factors that ultimately improve the design process include trust in the knowledge and expertise of other team members, visible and accessible design sharing among all team members, and strong social links inside and outside the team (Baird, Moore, and Jagodzinski, 2000).

Trust in the information source is an important criterion in engineers' selection of an information source. In formal communication sources such as journals, peer review is a built-in system that elicits trust in the authority and accuracy of the information content. But, since design engineers prefer oral communication sources, trust for them is derived from confidence in an individual's expertise. A study of engineering design teams at Rolls-Royce Aerospace found that face-to-face conversations with "a trusted engineer who understood the implications of the change on the tasks were seen to be significant to the rate of adoption" (Baird, Moore, and Jagodzinski, 2000). Engineers regularly give the name of their trusted sources of information when they communicate the new information to team members (and if they don't, the team members ask for the name).

Design teams rely on knowing what worked or did not work during similar projects undertaken by their organizations in the past. Again, the informal social networks are relied upon and experienced engineers are considered the most important source of information. Incorporating this type of trust and tacit knowledge into the organization's information system is difficult and is dependent on retaining experienced employees (Baird, Moore, and Jagodzinski, 2000).

Veshosky's survey of large engineering firms found that some had policies and procedures to assist engineers in locating information about innovations. These policies varied considerably, but included: assigning responsibility for information seeking to a specific group or individual; maintaining a file or database about information relating to innovations (including lessons learned on previous projects); preparing internal technical reports; conducting internal seminars; supporting internal libraries; encouraging participation in professional activities; and encouraging interaction with vendors.

The AIM-UK report makes recommendations for improving information services to academic and industry engineers. Recommendations fall into four broad categories:

1. Promoting improvements in information management at an organizational level
2. Promoting awareness of aerospace information resources in the public domain
3. Promoting access to aerospace information resources in the public domain
4. Expanding provision of training in research and information skills (Hanley, Harrington, and Blagden, 1998)

4.7 ENGINEERS' ADAPTATION TO INFORMATION INNOVATIONS

A survey of academic and industrial aerospace engineers in the United Kingdom in 1998 found a "depressing picture of low take up of [electronic] resources, particularly by industry." It revealed that electronic sources do not yet dominate in either sector, use of intranets is low, Internet use is lower in industry, and a third of academic respondents used the Internet seldom or never (Harrington and Blagden, 1999). The main reason for non-use "was a lack of awareness of these resources, with 70% of industrial respondents unaware of all resources and a staggering 50% of academic respondents: perhaps an indictment of university library user education" (Harrington and Blagden, 1999).

Many authors, particularly those writing in engineering journals or conference proceedings from engineering meetings, point out that engineers now use (or should use) the Internet, intranets, e-mail, the Web, and so forth to look for information, to share information, and to store or organize it (Sabharwal and Nicholson, 1997; Mathieu, 1995; Hallmark, 1995; Hoschette, 2000; Bender, et al., 1997). Others report on the utility (mostly anticipated utility) of internal electronic information management systems (Skinder and Gresehover, 1995; Yeaple, 1992).

Engineers use e-mail more than any other electronic communication channel, something that is not surprising given engineers' preference for informal and personal channels (Liebscher, Abels, and Denman, 1997). A 1994 study of engineering faculty in six small southeastern universities in the United States found that even in the mid-1990s, 86% of the faculty used computer networks. Of the users, 98% used e-mail, half at least daily, primarily to communicate with colleagues rather than with students. A majority used three or more network services, including subject-specific discussion groups and remote databases (Liebscher, Abels, and Denman). Later studies show a willingness to accept electronic journals but mostly as a convenience and supplement to print journals (Rusch-Feja and Siebeky, 1999; Meadows, 1997; Mehta and Young, 1995). More is said about use of electronic journals in Chapter 10.

Findings suggested that engineers are concerned about the move toward electronic information systems. In a study of 18 engineering companies in the United Kingdom, Harrington and

Bladgen (1999) found that all of the companies relied on a combination of manual and computer-based information systems. Almost all of the respondents expressed concerns, most commonly about: information overload and anxiety that something important will be missed; cultural problems of overcoming resistance to new technology; difficulty of data archiving and preservation; software obsolescence; and security.

One of the biggest advantages of the Web for engineers is the building of common-interest communities. This matches engineers' traditional information-seeking reliance on personal contacts. Access to other engineers and networking enhances many engineering projects because engineering expertise is increasingly geographically dispersed and interdisciplinary (Mathieu, 1995). In the engineering world, Engineering Information Village is an ambitious attempt to combine community building with a variety of engineering information resources. EI Village was redesigned based on user-feedback and has been described numerous times in the literature (Hollis, 1998; Tenopir, 1996; Bartenbach, 1996). In the United Kingdom, EEVL (Edinburgh Engineering Virtual Library) was an early attempt to build a gateway to high-quality Web resources (MacLeod and Kerr, 1997).

Integrated computerized knowledge management systems that capture and store tacit knowledge from engineering team members and other engineers in the organization, internally created written information, and external information sources are recommended to help ensure information is used effectively in engineering firms. Carstensen (1997) recommends that "a computer-based information exploration support system should provide easy access—including seamless switching between use of different search strategies and information sources"—to many types of information. Carstensen found that the most important types of information sources to be included in such a system include:

- Previous designs
- Design rationales
- Similar product information
- Known problems in products
- Component specifications
- Standards and norms
- Working procedures

- Production line characteristics
- Information on new materials and components
- Literature and research results
- Relevant persons
- Project documentation

Even documentation writing can be made more attractive to engineers if an organization uses cutting-edge software that makes documentation part of the design process from the start (Button and Sharrock, 1996). Such integrated information systems are not yet common in engineering firms, but since engineers in most countries now routinely use computer networks for a variety of work-related tasks (Pinelli, et al., 1997a, b), development of such systems should be a priority.

Information can be more effectively incorporated and used in the engineering workplace if engineers feel comfortable with a variety of information sources and are information literate. Cheuk (1998) studied information literacy among engineers in Singapore and concluded that, in the workplace, information literacy is required at both the individual and team-working level. Managers must recognize that information seeking and use is an iterative process and there is no one direct path to seeking information. Individuals must be encouraged to develop personal information strategies that are appropriate for different situations. Information needs are constantly changing and a successful information system in an engineering organization must provide access to a variety of information sources and systems and allow a flexible and individual approach to information search strategies. Cheuk cautions companies that "rigid guidelines based on the assumption that some information sources are more suitable for certain type of employees should be re-evaluated." An in-house study of information literacy levels within an organization can lead to the development of an information literacy strategy and curriculum that will ensure information resources are better incorporated into the engineering process (Cheuk, 1998).

Electronic journals and distribution of external information sources form a part of an integrated information system, but also offer advantages on their own. IEEE has studied the long-range effects of electronic publishing of their journals on their various participants (Herkert and Nielsen, 1998). Like many professional societies, IEEE is shifting to electronics as their primary distribution medium for their peer-reviewed scholarly journals and other publications.

A Delphi study commissioned in 1998 by IEEE revealed many obstacles that must be overcome before widespread adoption of electronic journals is possible. These include:

- *IEEE barriers:* IEEE must have a coherent vision, be a leader to avoid falling behind, set reasonable goals, and provide information in a flexible standard format.
- *Individual barriers:* A user-friendly interface and education and training must be provided to members.
- *Organizational and technological barriers:* Members must have access to hardware, software, and communications at a discounted rate; IEEE should develop organization-specific tools, have workstations in central locations like libraries, continue to provide print and electronic sources as long as demand exists, and provide financial incentives to switch (Herkert and Nielsen, 1998).

5

ENGINEERS' INFORMATION SEEKING AND USE

5.1 INTRODUCTION

Information seeking includes all of the processes an engineer goes through to look for, identify, and obtain relevant information. The research in these areas is uneven, with a decided emphasis on the looking for, rather than the obtaining. It is sometimes difficult to separate these tasks, however, as seeking patterns may change if the information is perceived as difficult to retrieve. Therefore, all parts of the process may be included under "seeking," but we have also highlighted some studies that specifically focus on information use through reading and interpersonal means such as listening. Perhaps no part of the communications process has been more studied than that of information seeking. The field of information science focuses more on studying information seeking than it does on information inputs through reading and listening. It concentrates even less on information outputs such as writing and speaking. There is a rich body of research literature that focuses specifically on the information seeking and input use of engineers; some of which we discuss in this chapter. In Chapter 6 we describe several factors that affect engineers' information seeking and input use.

Engineers' information seeking and use has remained an area of extensive study in many countries throughout the last several

Communication Patterns of Engineers. By Carol Tenopir and Donald W. King
ISBN 0-471-48492-X © 2004 Institute of Electrical and Electronics Engineers

decades. Yet many questions remain unanswered. The answers to these questions will not only provide compelling information about engineers, but they will also help information service providers develop better services for engineers and implement policies that help engineers more easily access quality information. Easy access is an engineer's top information priority, particularly for practitioners. Engineers often work under tight time restrictions and with proprietary information. These two factors limit the sources practicing engineers determine as valuable to their work. In academia, engineers tend to use a greater variety of resources and also tend to rely more on formal channels for information. Cost and the engineers' time are the greatest barriers to access. When engineers are unaware of the sources and services available to them, they cannot access high quality information efficiently. Again, in academia, engineers tend to be more aware of the services available through formal sources like libraries. Engineers need a large amount of information in order to initiate new projects and see them through successfully. Digital information resources help engineers to access quality information more efficiently, although many engineers have been reluctant to use such resources. Common-interest digital communities and well-designed internal information systems are beginning to improve information seeking and communication between engineers separated by geography and discipline.

In this chapter, we discuss the channels engineers use in seeking and using information. In section 5.3, we present evidence of the resources used in these input processes, including the time engineers expend in information seeking and use and their use of sources such as libraries and technologies. In Section 5.2, we describe the various information channels engineers use.

5.2 ENGINEERS' CHANNELS FOR INFORMATION SEEKING AND USE

Ellis and Haugan (1997) discovered eight categories of information seeking in their study of engineers and research scientists at Statoil, an international oil and gas company headquartered in Norway. These categories provide a general model for engineers' information seeking adapted from Ellis's earlier model (see Chapter 2). Information seeking falls into these eight categories:

1. *Surveying* (an initial search for an overview of the literature within a new subject field or to locate key people in the field)
2. *Chaining* (following chains of different forms of connection between sources to identify new sources of information, such as following citations in literature or references from individuals)
3. *Monitoring* (maintaining awareness of developments and technologies by regularly following specific formal or informal information sources)
4. *Browsing* (regular scanning of primary and secondary sources)
5. *Distinguishing* (ranking information sources according to their relative importance to the user)
6. *Filtering* (using search strategies that will make the information retrieved as relevant and precise as possible)
7. *Extracting* (working through sources to locate material of interest in those sources)
8. *Ending* (activities involved with finishing the information-seeking process, usually at the end of an R&D project)

The nature of the engineering workplace and the engineering discipline itself shapes the ways in which engineers seek information. Engineers typically work under deadlines and therefore seek immediate answers to specific questions, rather than a set of documents on a topic. They prefer easily accessible channels and sources and will choose the ones most easily accessed even over a higher-quality source (Toraki, 1999; Taylor, 1986). "Accessibility, perceived technical quality, and experience with the information channel or source are considered as the basic determinants for engineers in order to gain access to an information channel, following the law of least effort" (Toraki, 1999). Not surprisingly, libraries have never ranked high as a preferred source for information by most engineers.

Engineers' information-seeking channels tend to be either written or interpersonal (i.e., oral) in nature. The interpersonal input can involve informal discussions such as corridor talks, impromptu visits, cafeteria talks, telephone conversations, and so on, or it can involve more formal meetings such as presentations at conferences, attending classes, staff meetings, committee meetings, contractor meetings, brainstorming sessions, and so on. These interpersonal inputs may come from within the engineer's organization or external sources as inputs. Similarly, reading can involve informal e-mail, letters, memos, proposals, and so on, or more formal written materials such as scholarly articles, trade journals, professional books, internal and external reports, and patent documents.

Studies involving engineers done by King Research and the University of Tennessee from 1986 to 2001 yielded the estimates of engineers' reading activity given in Table 5.1.

Pinelli, et al. (1980) reported similar results for scholarly journal articles (80 readings), trade journals, bulletins (61 readings), and internal and external reports (104 readings). They also reported readings of letters (200), memos (292), proposals (34), conference or meeting papers (50), and drawing specifications (95), among others.

Journal articles are reported to be very important to engineers and scientists (Tenopir and King, 2000a), but they represent only a small fraction of the technical literature on most topics and may actually serve as an abstract of a wider body of literature. Technical reports, so-called "gray literature," are abundant. Esler and Nelson calculate that 123,000 technical reports are produced annually in federal labs, research universities, and corporate research labs. E-print services such as the arXiv.org system of Cornell University and the Department of Energy's PrePrint

Table 5.1   Average Annual Amount of Readings* by Engineers by Type of Document: U.S. 1986–2001

Type of Document	Readings	(%)
Scholarly journal articles	83	32.3
Trade journals, bulletins	47	18.3
Professional books	14	5.4
Other books	26	10.1
Internal reports	73	28.4
External reports	8	3.1
Patent documents	6	2.3
Total	257	99.9

*Readings are defined as going beyond the title and abstract to the body of the document. There can be multiple readings of a particular document.
Source: Surveys at: AT&T Bell Labs, Air Products & Chemicals, Inc., Baxter Healthcare, Eastman Chemical Co., Eastman Kodak Co., National Institutes of Health, Procter & Gamble Co., Oak Ridge National Laboratory, University of Tennessee ($n = 310$).

Network now make technical gray literature freely and widely available (Tenopir, et al., 2001; Lawal, 2002).

Since engineers favor informal channels and do-it-yourself information seeking, it is not surprising that personal communication remains the most popular channel for information. Engineers seek information from colleagues within their organizations, from clients, and from external experts and personal information collections. This finding has been reaffirmed over five decades of studies (Marquis and Allen, 1966; Allen, 1970; Shuchman, 1981; King, et al., 1994; Tenopir and King, 2000a).

Barriers to seeking both oral and written information were identified by Hertzum and Pejtersen. The major barrier to seeking written information was cost, and engineers greatly value their time. Therefore quick, easy access was identified to be "of paramount importance." Other barriers included irrelevant information, poor availability of information, unfriendly information-seeking tools, and too much intellectual effort required. Engineers reported that finding the right channels and sources among so much irrelevant information was difficult and a waste of effort.

A major barrier to oral information was cost in terms of time as well. The second most common barrier was "the intellectual and social effort required to present the information need in a way that triggers the other person's attention and gets him/her constructively involved." If the information request was not adequately presented, engineers often found the answer to be too general and not relevant to their specific problem. Additional barriers to seeking oral information were the need for confidential information, poor memory, and inappropriate information (Hertzum and Pejtersen, 2000).

In team-based projects and in all oral communication it is important is to be an effective listener (Kaye, 1998). Poor listening skills were found by Levitt and Howe (2000) to be the main reason engineering applicants failed in job interviews. Cerri (1999) recommends that engineers learn to understand the "human processes of perception, communication, and cognition" in a "7-Step Effective Communication Process." This process will help engineers become better listeners and to orally communicate more effectively (in other words, to learn to better understand oral inputs).

Cerri's seven steps are as follows:

1. Understand which of the five senses the listener is operating

and match it.

2. Build unconscious rapport by reducing listener's filtering by mirroring, matching, pacing, and leading verbal and nonverbal clues.
3. Uncover the listener's paradigms of reality.
4. Send the message.
5. Check to determine if it was received.
6. Go back to steps 1–3 if the message was not received.
7. Send the next message.

The process of receiving documents is no less complex than receiving oral messages.

Hertzum and Pejtersen (2000) found that personal contacts within engineers' workgroups and at conferences are also favored by engineers in Denmark, but, although they "display a strong preference for obtaining new information from people," documents are an important supporting source. Preferences for certain types of sources seem to be intertwined with information seeking patterns as engineers search through sources that are most readily available to them. Hertzum and Pejtersen conclude that engineers seem "somewhat biased toward getting information without deliberately searching for it. They spend time leafing through journals, talking to each other, attending conferences, and participating in other activities that subject them to a lot of information they had not consciously set out to look for." This bias is most common when they are looking for information about new developments on a topic new to them.

Rosenbloom and Wolek (1967) surveyed more than 3,000 engineers (and scientists) in large corporations and from a sample of members of the Institute of Electrical and Electronics Engineers. One principal focus of the data collection was to determine information channels used by engineers. Respondents were asked to report their most recent instance in which an item of information proved to be useful in their work (excluding someone in their immediate circle of colleagues). Channels used are summarized in Table 5.2.

Clearly, these engineers in the 1960s relied much more on channels found in their own organization than on external sources (63% versus 33%), and they relied more on interpersonal sources than on written materials (62% versus 43%).

Table 5.2 Channels Used by Engineers at the Institute of Electrical and Electronic Engineering

	Proportion of Instances (%)
Channels within own company	
Interpersonal	
Local source (within establishment)	25
Other corporate	26
Written media (documents)	12
Channels outside company	
Interpersonal (anyone outside company)	11
Written media	
Professional (books, articles, conference papers)	15
Trade (trade magazines, catalogs, technical reports)	11
	100

Source: Rosenbloom and Wolek (1970).

Allen reports comparisons observed in the early 1980s among information channels used in performing technological projects. Channels used in these projects are summarized in Table 5.3.

These results reinforce the notion that engineers are more dependent on colleagues than the literature.

5.3 RESOURCES USED BY ENGINEERS FOR INFORMATION SEEKING

The most important resource used in information seeking is engineers' time. The King Research studies in industry and government (1986–1998) showed that engineers spent about 26% of their time inputting information using various channels as shown in Table 5.4.

The reading time (280 hours) is estimated in Table 5.5:

It is clear that a considerable amount of engineers' time is spent in information seeking and use, and it is split evenly between interpersonal and reading channels. Raitt also provides some evidence of how much time is spent on various communication channels and sources, as shown in Table 5.6.

The interpersonal time seems to be more prominent than time spent using written channels.

**Table 5.3 Information Channels Used by
Engineers Performing Technological Projects**

	Proportion of Instances (%)
Literature	8
Vendors	14
Customer	19
Other external sources	9
Lab. technical staff	6
Company research programs	5
Analysis and experimentation	31
Previous personal experience	8
	100

Source: T. J. Allen (1988).

Allen, Shuchman, and Pinelli, et al. showed that engineers used library resources and librarians relatively frequently as a source for information for recent major projects. Pinelli, et al. (1997a,b) estimate that aerospace engineers use a library an average of 3.2 times per month (or about 38 times per year). Siess (1982) reports that libraries are used by engineers between 28 and 64 times a year depending on the type of research. King estimated in 1984 that engineers used a library an average of 54 times per year. For six organizations surveyed independently in the late 1980s and early

**Table 5.4 Average Annual Amount (Hours) and Proportion (%) of Time
Spent by Engineers in Seeking and Using Information by Type of
Channel: U.S. 1986–1998**

	Time Spent	
Type of Channel	Hours	%
Information input		
Informal discussions*	104	4.9
Attending internal meetings	136	6.4
Attending external meetings	34	1.6
Reading articles, reports, e-mail, etc.	280	13.1
Total Input	554	26.0

*Assumes that one-half of informal discussions is spent in receiving information and the other half in sending.
Source: Surveys at: AT&T Bell Labs, Air Products & Chemicals, Inc., Baxter Healthcare, Eastman Chemical Co., Eastman Kodak Co., National Institutes of Health, Procter & Gamble Co. (*n* = 252).

Table 5.5   Estimated Time Spent Reading by Engineers

Type of Document	Time Spent	
	Hours	%
Scholarly journal articles	72	25.7
Trade journals, bulletins	11	3.9
Professional books	19	6.8
Other books	24	8.6
Internal reports	58	20.7
External reports	6	2.1
Patent documents	4	1.4
Other (including email)	86	30.7
Total	280	99.9

*Assumes that one-half of informal discussions is spent in receiving information and the other half in sending.
Source: Surveys at: AT&T Bell Labs, Air Products & Chemicals, Inc., Baxter Healthcare, Eastman Chemical Co., Eastman Kodak Co., National Institutes of Health, Procter & Gamble Co. ($n = 252$).

Table 5.6   Proportion of Time Engineers Spend (Little Time, Quite a Lot of Time, or Very Much Time) on Various Communication Channels: European Aerospace 1984

Communication Channels	Level of Time (%)		
	Little	Quite a Lot	Very Much
Oral, formal			
Staff meetings	66	27	7
Contractor meetings	43	39	18
Presentations	77	22	1
Progress meetings	46	41	13
Brainstorming sessions	74	19	7
Committee meetings	77	19	4
Oral, informal			
Corridor talks	69	23	8
Canteen talks	84	15	1
Impromptu visits	38	39	24
Sports/social phone	38	40	22
Written			
Letter	50	41	9
Memo/telex	28	55	17
Internal report	36	49	15
Conference paper	77	19	4
External Paper/article	82	12	5
Giving documents	59	32	9

Source: Raitt ($n = 155$). Includes only those who answered the questions.

1990s, the average found was 39 uses per year, which is very close to Pinelli's observation. The 1984 estimate was from a random sample of engineers including academics, which may partially account for the difference between it and the surveys done in organizations in the late 1980s and early 1990s.

King and Griffiths performed independent, in-depth studies of library use in 31 organizations from 1982 to 1998. These studies show that libraries in organizations fill a very special niche in communication processes. For example, most older articles (more than about two years old) that are read by scientists and engineers come from libraries, and these articles are far more useful and valuable than articles read from personal subscriptions (because the latter are read most often for current awareness or continuing education purposes). Libraries are also used by engineers to read journals that are infrequently read by them and/or that are particularly expensive. Engineers and scientists generally act in an economically rational way when choosing from where they obtain literature. They take into account their time and journal prices. The substantial increase in journal prices over the years led engineers and scientists to decrease their number of personal journal subscriptions (5.8 per person in 1977, to 4.0 in 1984, to 3.7 in the late 1980s and early 1990s). The proportion of all readings that are from library-provided journals has increased: 18% in 1977, to 27% in 1984, and to 56% in the early 1990s. Even so, the number of personal subscriptions (about half of which are paid for by companies) far exceeds the number of library subscriptions in companies, typically by a ratio of 5 to 1. Griffiths and King (1993) demonstrated the usefulness and value of organization libraries and their services, and they cite a number of similar results reported by others.

Several in-depth studies have assessed Information Analysis Centers (IACs). A Coastal Engineering IAC was described by Weggel in 1973. Mason conducted a cost-benefit analysis in 1977. At about that time Corridore (1976) studied Department of Defense IACs, and Engineering Index, Inc. (1978) performed a study of IACs and numeric data provided by them. In the early 1980s, Roderer and King (1982) examined the use, usefulness, and value of two IACs: the Network Energy Software Center and the Radiation Shielding Information Center. Extensive studies have also examined federal clearinghouses, including a study by McClure, et al. (1986) on the National Technical Information Service and studies by Pinelli, et al. (1997a,b) concerning other federal centers.

Even when they do use a library, engineers like to search for information themselves rather than go through a librarian or other intermediary. In a study of users of the EPA library in Research Triangle Park, Grigg (1998) found that engineers were less likely than scientists to ask for a mediated online search. Although Kuhlthau's model (see Chapter 2) of information seeking found that mediation is most effective at the early exploration stage, engineers in Grigg's study hesitated to ask for assistance. "In fact, the tendency for engineers seems to be to do everything possible themselves until they reached a point where they had done all they could and still not found the needed information. Only at this point would engineers ask for a mediated search" (Grigg, 1998). Engineers told Grigg that the nature of their information needs makes it more effective to conduct their own searches. Even when they know a mediated search is needed, they like to conduct a preliminary search first. In those rare instances when engineers request a mediated online search, "they most likely used them only to act as a photocopying service for pre-identified articles" (Grigg, 1998).

Ignorance of libraries and library services is also part of the problem. A major study of United Kingdom aerospace engineers sponsored by the European Initiative in Libraries and Information in Aerospace (EURILIA) resulted in the AIM-UK report (Aerospace Information Management–UK). This report found that nearly half of the respondents considered their current information systems to be ineffective, but senior managers were not aware of the potential contributions that a librarian and library could make to their information management problems. Three quarters of industry respondents and 60% of academic respondents reported difficulty in identifying and locating useful materials on the Internet and many worried that they had missed significant information (Harrington and Blagden, 1999).

Some of the report's recommendations have been implemented since its publication in 1998, including a Knowledge Management Pilot Training Seminar for the aerospace industry and funding for a new aerospace and defense engineering Internet gateway at Cranfield University (Harrington and Blagden, 1999). The authors of the report recommended a national information policy in the United Kingdom that will promote a "national holistic view of how we as a nation can capitalize on our library and information resources and how we can increase information literacy (not information technology) across the nation" (Harrington and Blagden, 1999).

In 1995, as part of the United Kingdom's Electronic Libraries Programme, EEVL (Edinburgh Engineering Virtual Library) was founded as a gateway to engineering resources on the Internet. Such services provide easy access to selected relevant information and help engineers to locate high-quality resources from among the millions of questionable Internet resources more quickly (MacLeod and Kerr, 1997).

The European Union funded projects in the late 1990s to help improve information literacy for scientists and engineers under the umbrella EDUCATE (EnD User Courses in Information Access through Communication Technology) and spinoff projects such as DEDICATE (Distance Education Information Courses with Access Through Networks) to help train the trainers in libraries. A self-paced user education course called INTO INFO was one project developed for engineers and scientists along with professional development initiatives for librarians (Fjällbrant, 1977). Since technology changes quickly and because studies of members of the Special Libraries Association have found that only a small percentage of engineering librarians hold undergraduate degrees in science or engineering in addition to their graduate degrees in library and information sciences (Mosley, 1995), these programs for librarians are important.

Since information technology is advancing so rapidly, engineers have a growing need to keep up-to-date with technical developments and therefore require efficient techniques to learn of these developments. Gessesse (1994) pointed out that today's engineers cannot afford to remain ignorant of the myriad types of information resources available or to waste time struggling to find them. The librarian can save the engineer's time and the organization's money by teaching engineers how to locate relevant information resources.

Many academic engineers find their university library to be the most helpful information source and "showed a clear preference for consulting their own academic library to assist them with information problems in work-related situations" (Farah, 1993). Farah found that they return to a helpful provider because of prior use or experience (68%) and convenience (18%). When asked why they access a particular information provider, the faculty gave several reasons, including: accessibility (58%), quality/accuracy of information (51%), timeliness (35%), least costly in terms of time (29%), ease of use (24%), comprehensiveness (15%), referred to the source (8%), and least costly in terms of money (7%).

Colleges and universities can also influence how their engineering faculty use their networks, particularly by providing networked workstations in their offices. Training programs, including advanced training programs that target specific topics, will help them make more effective use of computer networks (Abels, Liebscher, and Denman, 1996).

Perhaps engineers do not use formal information systems adequately or efficiently, because current designs fail to reflect the way they think and the way they work. Yeaple (1992) recommends that database systems are easier for engineers to use if the systems "mirror an engineer's mental connections between chunks of information." Since an engineer makes mental connections around his or her functional and structural design hierarchies, systems that provide this structure are more easily comprehensible to engineers and reduce their information-seeking time.

Information technologies raise the communication requirements of engineers, including the need for up-to-date information from internal, governmental, and other sources and the need to share project details across workgroups and with clients. Sabharwal and Nicholson predict that the client side of this equation will increase in environmental engineering firms as clients become more sophisticated and "become more involved in decisions during project planning and design phases." Clients will begin to demand communication and interaction via the Internet and the Internet will make it easier for international companies to work together across geographic boundaries. They envision a future where large offices with high-paid executives and a large staff are no longer needed, as engineers work from home and Internet communication allows "closer, more immediate coordination between the environmental professional and the client."

Neither working engineers nor engineering students use physical libraries as much as scientists do—particularly if the libraries are not located nearby—and they overwhelmingly prefer digital information sources (Holland, 1998; Holland and Powell, 1995). Holland found that fewer than 20% of employees now use the corporate engineering library, preferring to rely on desktop information systems.

The hesitancy of engineers to use a physical library if it is not convenient to them does not necessarily translate to a preference for totally virtual libraries. As recently as the early 1990s, Pinelli found that aerospace engineers still favored "rather traditional information sources, such as personal information stores and discussions with colleagues, over newer electronic services" (Harrington and Blagden, 1999). In a study of engineers who belonged to the Technical Chamber of Greece (a professional organization that maintains a library for its members), most replied that "the virtual library might have a positive impact on their job." More than one-third of the respondents wanted a virtual library to replace the existing physical library, but the rest believed that personal contact and physical browsing were important (Toraki, 1999). European aerospace engineers reported increasing difficulty in identifying and locating materials that met their information needs; but their use of the Internet is increasing and they have a more positive attitude toward electronic access than in the recent past (O'Flaherty, 1997).

6

FACTORS AFFECTING INFORMATION SEEKING AND USE

6.1 INTRODUCTION

There are many factors that affect the ways engineers and scientists seek and use information. For example, some factors are distinctions based on demographic variables including: geographic or cultural differences; personal differences (such as gender, age, education, or experience); differences between branches of engineering; differences due to work role; and differences between engineers and scientists. Ready availability of technology is likely responsible for most of the impacts of geographic differences between engineers. Technology can affect the kind of information available, its quality, and the way in which that information can be used.

Few studies have examined gender differences in engineering. Those conducted found dissimilarities in self-confidence and information production between women and men. Women respondents were found to be less confident in their knowledge of sources and services and to produce more "secretarial" types of documents. An increase in information creation has been found to parallel increased age and experience. Information use and creation varies across engineering branches and work role although all engineers use both formal and informal channels and most

Communication Patterns of Engineers. By Carol Tenopir and Donald W. King
ISBN 0-471-48492-X © 2004 Institute of Electrical and Electronics Engineers

prefer to use informal channels. Engineers' information channels tend to be less open and more internal than scientists. Also, engineers tend to create less information than scientists because engineers are oriented toward the creation of technological products rather than documents. More research conducted on differences of information creation and use between engineers and between engineers and other disciplines would help information service providers better understand and meet the information needs of engineers in different contexts.

There are also many factors that affect engineers' information seeking and use, including:

- Geographic and cultural differences
- Branches of engineering
- Nature of work being performed
- Organization policies
- Personal characteristics

These and other factors are discussed in sections 6.2 through 6.5.

6.2 EFFECTS OF GEOGRAPHIC AND CULTURAL DIFFERENCES ON INFORMATION SEEKING AND USE

Engineers around the world are more alike than different. There is a unique body of knowledge, and some would say a unique way of communicating, that defines an engineer. Much of the discussion and conclusions in the previous chapters do not consider place and culture issues. Engineers spend much of their time communicating, albeit in less-than-optimal ways, whether located in the United States, United Kingdom, Denmark, Japan, India, or Saudi Arabia. Still, there is a rich body of literature which compared engineers in different countries or cultures, and differences did emerge.

One of the biggest differences is the availability of technology. Most engineers in the United States, the United Kingdom, and many other highly industrialized nations have had desktop access to computers and speedy telecommunications connections since at least the mid–1990s. Ready access means that Internet applications such as e-mail, listservs, and the Web are a routine part of these engineers' external communication patterns, and internal applications such as intranets, decision support systems, and computer-assisted design systems are also available. This is not yet true in other parts of the world.

Aerospace engineers across Western Europe "have similar information-seeking habits" and all are increasingly using the Internet. They also have a more positive attitude toward electronic access to information than in the past (O'Flaherty, 1997). Like their counterparts in the United States, European aerospace engineers are experiencing information overload, which creates problems "in making effective use of existing information services" (Harrington and Blagden, 1999).

There are many kinds of barriers to efficient information seeking; for example, engineers in India are not able to attend many professional conferences due to financial constraints (and engineering students in India are often not even aware of conferences) (Lalitha, 1995). In Saudi Arabia, university engineering faculty have significant difficulty acquiring all of the information they need, including delays in getting journals, outdated book collections, and lack of help in locating the information since libraries are not as well equipped as those in the U.K. (Al-Shanbari and Meadows, 1995). Although only 38% of the Saudi science and engineering faculty had a computer on their desk in the early 1990s, most of which were not networked, 80% had used a computer within the last year. Engineers used computers more than science faculty, with 93% of the engineering faculty claiming to make regular use of computers for teaching and research (Al-Shanbari and Meadows, 1995).

Russian scientists have experienced declining access to foreign scientific literature following the breakup of the Soviet Union. Russian scientists and engineers now rely on their western colleagues for up-to-date information and reprints of journal articles. Large serials budgets previously provided by the centralized Soviet system no longer exist. Technology is helping some, as private communications and article sharing via email have flourished with international colleagues (Markusova, et al., 1996).

Difficulties getting materials may have at least one unexpected positive consequence. Researchers who report the most difficulty getting materials also report that this encourages a high degree of collaboration both within their own country (Al-Shanbari and Meadows, 1995) and with western colleagues (Markusova, et al., 1996).

If technology access is optimized and equalized, the similarities across engineers around the world may far outweigh any cultural or geographic differences. When telecommunications problems are taken out of the equation, it was found that Saudi engineering faculty used CD-ROM technology equally with western faculty, although western faculty used e-mail more than Arabic faculty members (Al-Shanbari and Meadows, 1995). A study of university faculty in the United Kingdom and Czech Republic found differences in how they used paper information systems, but no differences in how they used electronic personal information management systems (Jones and Thomas, 1999).

In this era of multinational corporations and a global marketplace, engineers across national boundaries often work together on long-distance teams or find it advantageous to exchange information across national boundaries. According to Dimitrakis (1997), language difficulties and cultural differences are not the major impediments to successful communications between Asian and western engineers. Although there are many differences in the national cultures of Asia, such as Japan, Korea, Singapore, China, and Thailand, in all of these Dimitrakis believes "the most important criteria" for sharing information "are how sincere and knowledgeable the Asian engineers perceive you and how well you meet their expectations."

In general, Asian companies that rely on outside technologies are conservative and build their relationships on trust and a free exchange of technical information and expertise. There are differences in the type of information Asian companies want and how they make decisions. Dimitrakis pointed out that engineers in Japan want well-documented results presented according to timetables; those in Korea listen and are willing to take risks; engineers in Thailand value a responsive attitude more than technical content and may request information that may seem irrelevant to the specific question at hand. Personal characteristics that exist in any geographic area or culture can also affect how an engineer creates and uses information.

6.3 EFFECTS OF THE NATURE OF WORK ON INFORMATION SEEKING AND USE

6.3.1 Information Seeking within Engineering Discipline

Engineering is a diverse profession and there are some specific differences in communication patterns within the various subfields and branches as a result of this diversity. Thermal engineers, for example, use less mature technology and less formalized knowledge than, for example, stress engineers. Thermal engineers work more closely and have a more inward focus than other subfields (Atman, Bursic and Lozito, 1995). Computer science engineers spend less time writing than do other types of engineers (Kreth, 2000).

Computer and network use also vary across engineering subfields. Bishop found that aerospace engineers working in aerodynamics or flight dynamics and control are more likely to use networks than those in other branches of aerospace. Electrical, electronics, and computing engineers in Saudi Arabia were found to use computers more than other types of engineers (Al-Shanbari and Meadows, 1995).

Construction engineering requires good communication between several parties with different yet vested interests. For example, the user *must* communicate his or her needs to a developer/architect. Then, once plans have been developed and a bid has been taken, the general contractor must understand the plans and *communicate* the plans to the subcontractors. The project-managing engineers will be the ones who ultimately make the installation decisions based on the communications from user to architect to general contracting to subcontracting and also based on electrical plans for the facility (Mench, 2002).

6.3.2 Differences in Information Seeking Due to Work Role

Engineers, typically those in large organizations, hold a variety of jobs that require a variety of tasks. Information needs also vary, depending on the nature of the work. For example, a research engineer needs theoretical information, while a design engineer needs information about existing materials, devices, and systems specific to a particular situation (Gessesse, 1994).

R&D and design engineers have received the most scholarly attention and seem to display distinct information patterns. Kim (1998) discusses many studies that found R&D engineers differ from other engineers in that they prefer problem solving through oral communication, spend less time on formal information sources, and use textbooks and reports more than journals. Vest, Long, and Anderson found that communication patterns of R&D engineers "remain distinct from those of other engineers." In their study they found that R&D engineers spend less time on communication (47% versus 62% for non-R&D engineers) and, when they are communicating, spend less time communicating with those outside their own group (18% for R&D engineers versus 25% for non-R&D engineers) and have a smaller communication network (6 or fewer people daily for R&D engineers versus more than 12 for non-R&D engineers) (Vest, Long and Anderson, 1996).

Designers need to communicate with people with specific competencies and previous experience as information resources. They need to explain, discuss, and argue issues during the design process. Even when they search for documents, they use them to find people to use as sources of oral information or to find people who can send them documents (Hertzum and Pejtersen, 2000). Design engineers rely on working with a team of specialists and therefore they coordinate knowledge. Court, Culley and McMahon (1997) found that this sharing of information is accomplished not only through face-to-face meetings, but also through use of video conferencing, e-mail, memos, formal documents, and so forth. The selection of communication medium influences the "richness of information that can be processed." Unfortunately, design team members seldom receive any formal training or guidance on which channels are appropriate for which tasks. The various communication channels have different levels of richness, as illustrated in Figure 6.1, presented by Court, Culley, and McMahon (1997). Face-to-face is the richest, most intense, most immediate communication mode, revealing multiple clues to the communicators.

Design engineers need information throughout the design

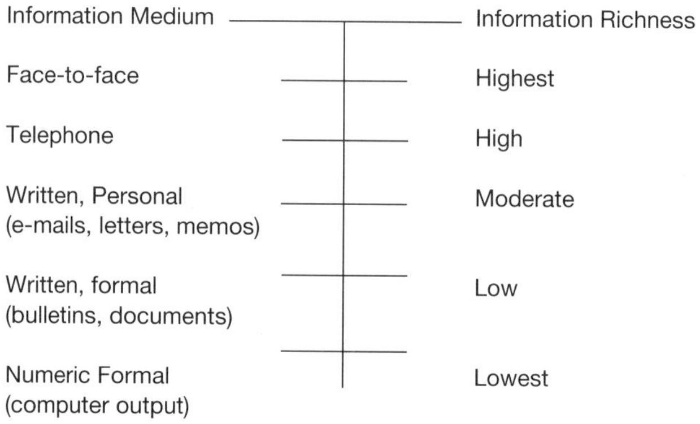

Information Medium	Information Richness
Face-to-face	Highest
Telephone	High
Written, Personal (e-mails, letters, memos)	Moderate
Written, formal (bulletins, documents)	Low
Numeric Formal (computer output)	Lowest

Figure 6.1 Communication Channel and Information. *Source:* Court, Culley, and McMahon (1997).

process, but they need different types of information depending on the type of design activity and stage of the design. Court, Culley and McMahon (1995) described nine types of designs (simple designs, complex designs, original designs, adaptive designs, variant designs, new designs, previously done designs, design tasks, and selection tasks) within three general categories: original, adaptive, and variant. For all types of design tasks engineers prefer personal information sources including memory, colleagues, and other employees. Engineers access these sources via sight, sound, and speech via telephones or speaking in-person. Even when accessing external information they use personal-access paths.

Design engineers in the United Kingdom preferred to follow well-established paths to seek and locate information and concentrated on sources and methods that were familiar to them. Personal memory was relied on 30% of the time, which may pose problems to companies when employees change jobs (Court, Culley, and McMahon, 1995).

Design teams use information to make decisions that impact their entire organization. Their role has been complicated recently "by the ever increasing amounts of information that are being produced by specialist groups and organizations, and by the wide variety of formats and media in which they are delivered and presented . . . the sheer volume may slow down or prevent the engineer or designer obtaining a critical fact or piece of information" (Court, Culley and McMahon, 1997).

A survey of 200 engineering designers in the United Kingdom and 20 case studies revealed a wide range of written channels. Internal channels included: product specifications, previous design schemes, existing design reports, other department reports, data handbooks, development and test data, sales data, commercial data, marketing data, manufacturing data, in-house parts catalogues, design guides, and service feedback. External sources included: journals, magazines, catalogs, libraries, professional organizations, academic institutions, the government, design guides, and events (Court, Culley and McMahon, 1997). Hertzum and Pejtersen (2000) also found that engineers make extensive use not only of interpersonal communications in their day-to-day work but also of information found in documents such as handbooks and internal reports.

Design engineering requires much cooperation and cooperative revisions. It has been described as a social activity, rather than a technical one, due to this high degree of interactivity (Lloyd, 2000). Good groupwork software assists design engineers more than others because of the nature of design work (Lee and Decker, 1994).

Information habits of the same individual may vary according to the complexity of the task underway. Byström and Järvelin (1995) classified tasks into five categories, the complexity of each influencing information seeking. These tasks, in order from most to least complex, are:

1. *Genuine decision task* (unexpected, new, unstructured; neither results, process, nor information requirements are known; for example, the collapse of the Soviet Union from the viewpoint of other governments).

2. *Known genuine decision task* (type and structure of result is known, but permanent procedures haven't yet emerged; for example, deciding the location of a new factory).

3. *Normal decision task* (structured, but requires case-based arbitration; for example, grading a student's paper).

4. *Normal information processing task* (mostly determinable, but requires some case-based arbitration; for example, the interpretation of a tax code).

5. *Automatic information-processing tasks* (completely determinable; for example, the computation of a person's net salary).

The number of sources consulted and the number of external sources used increased with the complexity of the task, but the use of internal channels and success of information seeking decreased.

Even in the same branch of engineering or the same company, an engineer may find that his or her work role changes over time. Typically, as an individual advances in seniority and rank, that individual assumes more administrative or managerial duties and fewer product design or pure engineering duties. Throughout a career an individual engineer may assume a variety of functions, "including research and development, design, testing, manufacturing and construction, sales, consulting, government and management, and teaching" (Leckie, Pettigrew, and Sylvain, 1996; Kemper, 1990). These changes in work role influence the type of information needed, the way information is sought, and the types of output expected. Design engineers want original, up-to-date information, relying heavily on internal reports and test results rather than the published literature. In a consulting role they rely more on external market information about vendors and customers. When an engineer takes on an administrative role, he or she needs a wider variety of both external and internal information, including regulations, information on new technologies, personnel records, and business information. R&D information needs similarly vary with each stage of the project (Leckie, Pettigrew, and Sylvain, 1990).

Engineers in the workplace often have different communication patterns than those in academia. Part of this is due to differing roles: the focus on research and teaching in academia versus the focus on production in industry. Network use is higher in academia and use increases with the amount of education. Engineers who have teaching and research responsibilities tend to use computer networks more (Bishop, 1994). Academics read more journal articles and write much more than scientists or engineers in industry (Tenopir and King, 2000a). However, aerospace engineers working in academic institutions and those working in industry are found to have similar information-seeking habits (O'Flaherty, 1997).

Textual information is not the only type of written information needed by engineers. A variety of graphical information, including plans, graphs, maps, structural layouts and so on. are neccessary for many engineering applications. An offshore-drilling platform, for example, requires 600,000 engineering graphical documents (Rodriguez et al., 1994). These graphical documents pose many unique access and storage problems, including the physical management of paper-based materials and the massive amounts of storage space required for digital versions.

Sometimes the internal colleague who assists the information-seeking process is an individual in the organization variously called a "gatekeeper," "boundary-spanner," "high performer," or "hunter-gatherer" (Sonnenwald and Pierce, 2000; Kim, 1998; Baird, Moore and Jagodzinski, 2000). The gatekeepers are more oriented toward external information sources than the typical engineer and are often called on by others in the organization for information (Sonnenwald and Pierce, 2000; Kim, 1998). They read more, conduct more online searches, make more formal presentations, and publish more articles than the average engineer (Sonnenwald and Pierce, 2000; Kim, 2000; Tenopir and King, 2000a). Gatekeepers are scientists or engineers who know information sources, have many contacts, and informally assist others in the organization to seek for and locate all types of information. They read, search online, present, and publish more papers than their colleagues and are more familiar with external sources. As with other sources of information that are valued by engineers, trust is key. Trust in an individual's knowledge and expertise is why his or her expertise is sought. Young engineers whose supervisors were gatekeepers were found to have a lower rate of turnover and were promoted more frequently (Katz and Tushman, 1979; Lee, 1994). Young engineers fear they will hurt their reputations with their peers if they ask for the information they need, so having a supervisor that will help them find what they need seems to have a direct effect on their success.

In academia, engineering faculty and students rely more on formal communication channels and use a broad range of information resources, including books, journals, and technical reports. The library is an important source for these resources and databases and other finding aids are needed to locate them (Dessouky, 1994). Farah studied computer-engineering faculty in eight universities. Just as with industrial engineers, internal colleagues are sources for information (62%). Colleagues at professional meetings and other colleagues outside of their institution are also frequent resources. Unlike industrial engineers, however, the most frequently used information source was the university library (78%). Academic libraries within their institution were most frequently cited as the most helpful information providers for these engineers (Farah, 1993).

6.3.3 Additional Work-Related Factors Affecting Information Seeking

Information-seeking patterns vary with the type of R&D projects. The study of an R&D department by Ellis and Haugan (1997) identified three distinct types of projects that influence patterns of information seeking: For *incremental projects,* face-to-face communication was the preferred mode of information seeking, particularly with in-house personnel. They select information channels by asking others, through personal knowledge and experience, and by consulting the library. The engineers need both broad knowledge and project-oriented in-depth knowledge for problem solving, looking for particular equipment or methods, comparing alternative solutions, and seeking ideas for their own experiments.

For *radical projects* (those that go beyond existing knowledge) R&D workers preferred communicating with more experienced colleagues, contractors, and suppliers. They preferred internal communication channels and used these channels before they used databases to search the literature. However, they did turn to published literature because "published literature is recognized as the easiest means to get information and is often chosen 'to take the line of least resistance.'" Review articles and conference proceedings were first. Conference proceedings were considered poor quality and out-of-date, however, when compared with scientific journals. Engineers used scientific journals to keep up with current developments, but they were considered to be time consuming to use and too out-of-date. Review articles were used for interdisciplinary work and were often combined with online database searches. Ellis and Haugan also found that different types of materials were sought depending on the type of radical project. Research reports gave an overview of a field and internal information provided the commercial angle on such things as pricing (Ellis and Haugan, 1997).

Fundamental projects enhanced a company's understanding and provided a long-term perspective. A literature search was usually the first approach in these projects because the topics may be unfamiliar to the researcher. A library is the primary source for this type of project followed by the researcher's own knowledge and experience. The information-seeking process in these projects is highly interactive and includes both current awareness and retrospective searching (Ellis and Haugan, 1997).

Ellis and Haugan found that information seeking was "most extensive in the initial phase of a project, when both formal and informal channels" were used. In advanced stages, researchers were more selective and specific as they learned more about the problem. They observed that in the middle phases of a project the use of formal channels decreased and person-to-person communication dominated. This changed again in the final phase when both formal and informal channels were utilized, "mainly in the form of a small literature search or through contacts with knowledgeable persons in the field to supplement the information already gathered. Information gathering is an iterative process, and recurring activities often take place when new situations occur over the life of projects" (Ellis and Haugan, 1997). Other studies have found slightly different results.

R&D researchers need the most information at the start of a research project as new ideas and new technologies are explored. Hirsh reports that researchers need business and technical information at this early stage, which make external information resources more valuable, including journals, the Web, patent databases, business and news resources, and consultations with librarians. They read more and spend more time sending and receiving email at this first stage of research (Hirsh, 1999).

The second stage of R&D projects is the development phase, when the focus shifts from exploring new ideas to making technical breakthroughs. Hirsh (1999) found patent literature, technical journal articles, and conference attendance were most valuable at this stage.

After the development phase comes the transfer stage when a product plan is developed with internal partners. At this stage, in addition to technical journals and patents, researchers and managers use internal information, vendor information, textbooks/technical books, conference papers, interactions with colleagues, and personal sources of information. At any research stage participants spend little time on average searching the Internet or an intranet for work-related information, but 38% used the Web (Hirsh, 1999).

Von Seggern and Jourdain (1996) found the priority order of information sources used in problem solving to be: personal store of technical information, speaking with a co-worker or others inside the organization, speaking with colleagues outside the organization, literature sources in the organization's library, searching an electronic database in the library, and speaking with a librarian or technical information specialist.

Veshosky (1998) surveyed organizational and project managers at top engineering firms in the United States to measure their attitudes toward innovation and the importance of information for innovation. Project managers relied mostly on conversations with colleagues within their workgroup both to keep up-to-date and to help solve problems. Trade magazines were more important for keeping up-to-date than for problem solving, and conversations with external people (including clients, vendors, and colleagues) were more important for problem solving.

The value and choice of specific types of sources is not static. Veshosky found variation during the problem-solving process. In the early stages of the process, project managers favor "conversations with internal colleagues, consultants, subcontractors, clients, and vendors and review of personal files." In later stages, they use textbooks and handbooks, codes and standards, industry newsletters, and discussions with academic researchers (Veshosky, 1998). Figure 6.2 shows different sources throughout the stages of the information-seeking process of innovating.

Some resources are rarely, if ever, used for problem solving. These include professional conferences, short courses, regular academic courses, or conference proceedings. Other than conversations with various internal and external sources (which hold the top four places), information sources used by the highest percentage of respondents "regularly or occasionally for problem solving" are (in rank order): product literature, internal technical lectures, internal technical reports, industry technical reports, trade magazines, and professional meetings or seminars. Demonstrations were occasionally used by 55% of respondents, but 32% never used them (Veshosky, 1998).

Information seeking varied with the complexity of the task, as described by Järvelin. Figure 6.3 presents five categories of tasks, ranging in complexity and information-seeking patterns.

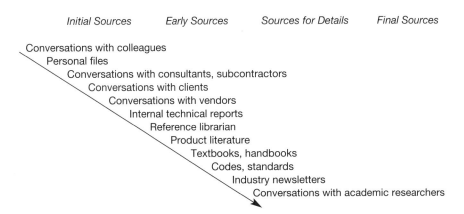

Initial Sources   *Early Sources*   *Sources for Details*   *Final Sources*

Conversations with colleagues
Personal files
Conversations with consultants, subcontractors
Conversations with clients
Conversations with vendors
Internal technical reports
Reference librarian
Product literature
Textbooks, handbooks
Codes, standards
Industry newsletters
Conversations with academic researchers

Figure 6.2 Information-Seeking Behavior During the Problem-Solving Process. *Source:* Veshosky (1998).

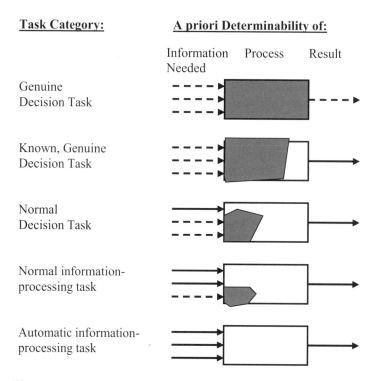

Figure 6.3   Task Categories. *Source:* Byström and Järvelin (1995).

6.4   EFFECTS OF ORGANIZATION POLICIES ON INFORMATION SEEKING AND USE

Veshosky surveyed large engineering firms and found that some had policies and procedures to assist engineers in locating information on innovations. These policies varied considerably, but included: assigning responsibility for information seeking to a specific group or individual; maintaining a file or database about information relating to innovations (including lessons learned on previous projects); preparing internal technical reports; conducting internal seminars; supporting internal libraries; encouraging participation in professional activities; and encouraging interaction with vendors. One approach to improving engineering communications is to connect engineers who have created new technologies with information users who are unfamiliar with the technologies. An approach used by C. P. Snow to connect the sciences and humanities is the basis for this approach (Amare, 2000).

Engineers and project managers tended to focus their efforts on obtaining internal information only. Unfortunately many were found to be unaware of the procedures within their organizations. These procedures might have helped in obtaining valuable external information. Misguided policies can also adversely affect information seeking. Hertzum and Pejtersen (2000) found that companies recommended that engineers search for information first in internal corporate archives to save time and money. But the engineers were reluctant to search there unless they were looking for specific materials that they already knew were in the archives. Even then, some preferred to go directly to the original author of the material and found this approach faster and more rewarding. None of the interviewed engineers often used the next step in information seeking—searching for external sources either on their own or through the library. Recognizing that engineers do not make adequate use of online services for continuing professional development, a new service was developed in Wales to support an integrated online learning environment (Lloyds, Moore, and Kitching, 2001).

6.5   EFFECTS OF PERSONAL CHARACTERISTICS ON INFORMATION SEEKING AND USE

Personal characteristics cause variation in information seeking, use, and creation. The most obvious is gender, but few studies take gender into account as a possible variable in engineers' communication patterns. Perhaps this is because engineering is still a male-dominated profession. Women are a small percentage of the 320,000 professionals who are members of IEEE. This is unlikely to change any time soon, because only approximately 20% of engineering students who graduated with a bachelor's degree in 1999 were women. Additional factors reported in this chapter are thought to have more influence on job-related communication patterns than gender, but some studies have examined gender as a variable.

A recent study by the Goodman Research Group, Inc. (2002) suggests that gender disparity in engineering disciplines, as opposed to fields such as medicine or law, is related to a lack of social support and mentorship opportunities available to female engineering students. The Goodman report found that female engineering students who dropped out had left their programs even though they performed well academically. Mentorship is an important element in engineers' informal communication networks. Goodman indicated that women, as students and professional engineers, need informal networks of communication and support as much as or more than male engineers.

Kreth (2000) surveyed recent engineering graduates of an unnamed midwestern university to study their writing experiences in industry and in coursework. She found gender differences, although her responses were more evenly split (85 men and 77 women) than the population of engineers. She cautions that many of the differences found may have more to do with branch of engineering than gender. For example, very few female respondents were electronics engineers, and electronics engineers are more likely than other engineers to write specifications. Kreth found that female industrial engineers spent the most time writing, followed by male mechanical and male industrial engineers. There were also differences in preferred methods for improving their writing. Women preferred sample documents and practice writing in technical writing courses; men preferred feedback from co-workers or supervisors.

The most disturbing aspect of Kreth's findings was the fact her results "suggest that women students were doing more secretarial types of writing tasks than their male counterparts." She found that a higher percentage of women wrote meeting notes (43% of the women versus 32% of the men), memos (64% of the women versus 57% of the men), and e-mail (40% of the women versus 33% of the men). However, she again cautions, "it is not clear that these differences are solely accountable for by gender" (Kreth, 2000). In addition, a higher percentage of men wrote so-called "prestige" documents, such as reports (71% of the men versus 53% of the women), specifications (39% of the men versus 25% of the women), and proposals (26% of the men versus 23% of the women). Ercegovac (1988) studied gender differences among engineering students' information literacy skills at UCLA. She found male students more confident than female students in their knowledge of engineering sources and services.

Some studies have found that older engineers are slightly less likely to use computers or networks; however, this trend may have changed and is not unique to engineering (Al-Shanbari and Meadows, 1995; Bishop, 1994). Others found that age is not a factor in adoption of electronic journals, although status of the individual may be (Pullinger, and Baldwin, 2002; Mahé, Andrys and Chartron, 2000). Gupta, et al. (1981) found that research productivity among Indian scientists, as measured by number of papers, books, or technical reports published, peaks when a scientist reaches between 26 and 30 years of experience, when most researchers are in their late fifties, and then tapers off. In contrast, productivity measured in terms of number of patents continues to increase with professional age. Results in an earlier study by Lemoine (1991) differed in that he found the number of patents decreased with the age of the researcher, while journal articles did not.

As an engineer matures and job responsibilities change, so may the types of information sources needed. Engineers who have been working in the same area for a long time rely on browsing known sources such as relevant journals and conference proceedings to stay current (Hertzum and Pejtersen, 2000). As they move into more business and managerial responsibilities they read more business newspapers and magazines (Holland and Powell, 1995).

7

INFORMATION OUTPUT
BY ENGINEERS

7.1 INTRODUCTION

In Chapter 4 we discussed the engineering profession and the nature of engineers' work. In Chapter 5 we described the information inputs to this work and the extent to which engineers sought and used information. Chapter 6 discussed factors that affect engineers' information seeking and use. In this chapter we describe the information output from engineers' work. This involves both interpersonal (i.e., oral) and written channels of communication. Interpersonal communication may involve such channels as informal discussions; providing substantive advice or consultation; research and other presentations; proposals or plans; and formal workshops, seminars, or university classes. Channels may target audiences that are either internal or external to the engineers' organization; however, engineers tend to communicate more internally than externally. Written communication channels include formal publications intended for external audiences such as scholarly and trade articles, conference proceedings, books, and patent documents. Technical reports, proposals, or plans are written for either internal or external consumption.

Engineers devote a substantial amount of time to communicating through interpersonal and written channels. In fact, from

Communication Patterns of Engineers. By Carol Tenopir and Donald W. King
ISBN 0-471-48492-X © 2004 Institute of Electrical and Electronics Engineers

King Research surveys, we estimate that engineers in industry and government devote about 690 hours to communicating information outputs. It is interesting to note that engineers spend more time communicating information output than inputting information. That is, engineers spend 690 hours per year in information output versus about 550 hours in information input. Pinelli, et al. (1989) reported evidence that aeronautical engineers spend 13.95 hours per week communicating technical information to others and 12.57 hours working with technical communications received from others, supporting to the above assertion.

Obviously one purpose of formally recording (or orally communicating) the output of one's work is to provide evidence of accomplishment; however, there are other consequences that undoubtedly far outweigh that purpose. One reason is to gain feedback from others, and this is particularly well achieved by talking. A longer-range beneficial consequence of this is the consumption and use of information by others. Thus, professionals communicate the results of their work for a variety of purposes in different ways (internal, external, formal, informal). Documenting one's research is an essential part of research in that the discipline of writing is, in itself, part of the creative process and often leads to new ideas. Documentation also satisfies researchers' needs for recognition and should record their research accomplishments.

Substantial evidence suggests that today's engineers spend a large proportion of their time communicating information output from their work and that good writing and oral communication skills are critical for success. Products and processes of engineering are communicated at every stage of the engineering cycle and effectively communicating may determine success and failure. In fact, Gunn (1995) of Michigan State University argues that "without communication there is not engineering." Levitt and Howe (2000) claim communication skills are as important as technical, analytical, and problem-solving skills "because information becomes knowledge only when conclusions drawn from analyses and/or potential solutions to problems are communicated to those who need to make decisions or implement solutions." Engineers, like all technical professionals, can accomplish more if they communicate effectively (Cerri, 1999). Engineering educators are increasingly aware that good communication skills must be an integral part of today's engineering curricula (Williams, 2002).

In this chapter we describe the channels used by engineers to communicate the information output of their work and the resources and tools used by them. We also discuss written communication channels in greater detail, describing their importance, problems observed in writing, and some solutions. In Chapter 8 we discuss ways in which education can change to enhance the vital communication processes of engineers.

7.2   CHANNELS ENGINEERS USE TO COMMUNICATE INFORMATION OUTPUT

Engineers use several channels to communicate information. Table 7.1 gives estimates of the extent to which these channels are used by industry and government engineers involved in surveys by King Research.

Pinelli, et al. (1988) observed similar outputs for aeronautical engineers they surveyed: technical reports (7.0 reports vs. 5.1 above), proposals (3.6 vs. 2.2), specifications (6.4 vs. 5.2), computer program documentation (2.6 vs. 2.5), scholarly and trade journal articles (1.4 vs. 0.1), conference proceedings (2.2 vs. 0.5), and memos (57.6 vs. 19.0). The latter three outputs may be higher in Pinelli's surveys because university engineers are included.

Technology, globalization, and the growth of interdisciplinary projects are changing the engineering workplace. These changes impact the expectations for communication by engineers, with a variety of oral and written outputs now a critical part of all types of engineering jobs. Few would disagree with Burdan and Strother (1995), who claim that "communication skills have become an essential element for employability in today's business world," regardless of the job area. A survey of 67 companies and government agencies in 1994 found that good communication skills were among the top six most important attributes looked for in recruiting and hiring (Miller and Olds, 1994). Blake (1998) believes that billions of dollars are lost in terms of corporate productivity and profitability yearly when engineers have problems with written communication.

Many agree that many, if not most, engineers have trouble writing and speaking clearly. A survey by Mahan, et al. (2000) found that industry is looking for good communicators, but many engineers are unable to present ideas clearly, describe the reason behind communicating, or link sentences into logical paragraphs. They reported long, rambling reports that are not well organized and relied on too much jargon. Robar (1998) reported on common

Table 7.1   Average Annual Output by Industry and Government Engineers by Type of Output: Observed, 1986–1996

Type of Output	Amount of Output
Oral Communication	
Substantive consultation, advice (times given)	240
Presentations about research and other work (no. of presentations)	
Internal meetings	4.0
External meetings	0.3
Presentations about proposals/plans	
Internal meetings	15.9
External meetings	11.1
Workshops, seminars, university classes	
Internal meetings	0.4
External meetings, classes	0.3
Written Communication	
Scholarly journal articles (no. of articles)	
Sole author	0.01
Co-author	0.04
Books (no. of books)	
Sole author	0.001
Co-author	0.003
Patent documents (over 5 years) (no. of patents)	
Application sole author	0.25
Patent granted sole author	0.07
Application co-author	0.65
Patent granted co-author	0.19
Other publications (e.g., conference proceeding)	
Sole author	0.11
Co-author	0.60
Technical reports (no. of reports)	5.1
Specifications (no. of specifications)	5.2
Computer documentation	2.8
Written proposals, plans (no. of proposals)	2.2
Memos	19.0
Other Communication	
Publications edited, reviewed, refereed	0.9
Contacts with suppliers, vendors, etc.	
Letter	1.7
Telephone	6.2
Visit	1.8

Source: Surveys at: AT&T Bell Labs, Air Products & Chemicals, Inc., Baxter Healthcare, Eastman Chemical Co., Eastman Kodak Co., National Institutes of Health, Procter & Gamble Co. ($n = 252$).

barriers to communication by engineers, including: lack of knowledge or experience; poorly defined ideas; messy written communications; one-sided or inappropriate communication; failure to bridge differences in values, attitudes, or perceptions with the audience; and poor listening skills. Although well-written reports, feasibility studies, memos, and letters enhance the image of a firm, engineers are rarely taught how to write well, let alone styles of writing for specific document types (Schillaci, 1996).

The engineering research process may produce multiple versions of an idea, each of which may be in different formats, media, and levels of detail, and each requiring a different type of oral or written expertise and style. Written and oral communication may be in informal or formal modes. According to Ercegovac (1988) , an original idea goes through distinct developmental stages, perhaps starting at an informal presentation in a university class, where nothing is written. As the idea matures, the researcher may decide to present the idea as a "work-in-progress" at a conference. Next, the idea may be conceptualized into a short paper or case study or pilot study. This paper will be included in conference proceedings, indexed and, for the first time, a bibliographic record of the title, authors, affiliations, and so forth. will be recorded. The work then progresses through other technical formats, "each having distinct characteristics and access mechanisms." These other formats may include technical reports, conference papers, dissertations, journal articles, chapters in books, patents, books, and articles in encyclopedias (Ercegovac, 1988). Various levels and types of oral communication may be added to each stage in Ercegovac's model, including communicating the idea in face-to-face meetings, over the telephone, and so on (see also Figures 2.1. and 2.2).

Communal writing is a feature of engineering design teams. In R&D or design situations, multiple forms of communication are particularly important for project success. Kim (1998) reported many studies over the years that found that "high performers on research projects had extensive communication with project team members." Teamwork depends on good interpersonal skills, as does the ability to talk to customers and clients (Owen, 1999).

Engineers who can communicate well are more likely to find employment, be promoted, and advance throughout their careers (Robar, 1998). According to Robinson (2000) the "new model" of engineers who have better communication skills are "winning the top jobs in high-tech firms." Similarly, managers who communicate well help reduce stress among the staff (Guindon, 1994). Surveys in many countries over the years found productive research scientists and engineers are more involved in communication than their less-productive peers (Al-Shanbari and Meadows, 1995; Lufkin and Miller, 1966; Griffiths and King, 1993; Tenopir and King, 2000a).

7.3 RESOURCES AND TOOLS ENGINEERS USE FOR COMMUNICATING INFORMATION OUTPUT

As with information input, the most important resource used by engineers to communicate the information output from their work is their time. Surveys by King Research involving engineers in industry and government suggest that they spend substantial time communicating through various channels.

Engineers in industry and government are estimated to spend nearly one-third of their time communicating the information out-

Table 7.2 **Average Annual Amount (Hours) and Proportion (%) of Time Spent by Engineers in Communicating the Information Output from Their Work by Channels Used: U.S. 1986–1998**

	Time Spent	
Type of Channel	Hours	(%)**
Informal discussions*	104	4.9
Consulting/advising others	222	10.4
Internal presentations	99	4.7
External presentations	24	1.1
Writing		
• proposals and plans	92	4.3
• technical reports	117	5.5
• articles, books, etc.	12	0.6
• programs, software	17	0.7
Total Output	687	32.2

*Assumes that one-half of informal discussions (52 hours) is spent in communicating information and the other half in receiving.

**Proportion of time spent is based on an average of 2,128 hours spent annually by engineers on work-related activities.

Source: Surveys at: AT&T Bell Labs, Air Products & Chemicals, Inc., Baxter Healthcare, Eastman Chemical Co., Eastman Kodak Co., National Institutes of Health, Procter & Gamble Co. (*n* = 252).

put from their work and about one-third of this time is spent writing. That is, much of their valuable time is spent writing, making this an important engineering activity that should not be ignored. Engineers use other resources for communicating, such as support staff and equipment (e.g., word processors, printers, photocopiers, projectors, video, telephones, CDs, etc.).

E-mail has reinforced the need for good writing skills (Mahan, et al., 2000) and face-to-face meetings are being replaced, or at least enhanced, with information technology (Skeris, 1998). E-mail is increasingly used to overcome "the limitations of face-to-face communication; to communicate with others in distant cities; to distribute general and technical information to a large number of people"; to discuss complex problems; to help structure arguments; to refine and distill a message; and to create an electronic paper-trail (Zimmerman, et al., 1994).

Many authors, particularly those writing in engineering journals or conference proceedings, point out that engineers now use (or should use) the Internet, intranets, e-mail, the Web, and so forth, to look for information, to share information, and to store or organize it (Sabharwal and Nicholson, 1997; Mathieu, 1995; Hallmark, 1995; Hoschette, 2000; Bender, et al., 1997). Others report on the utility, albeit anticipated utility, of internal electronic information management systems for engineering companies (Skinder and Gresehover, 1995; Yeaple, 1992).

Lack of standardization of data exchange also implies effective communication, especially between engineering teams that are geographically dispersed, as well as among others who might use different systems. Models for data exchange developed under the SEDRES project (Johnson, et al., 1999; Herzog and Törne, 1999; Harris and Candy, 1999) show that standardized information models increase the ease and speed with which information is shared between engineers. This project may serve as an example to be emulated. However, technological solutions cannot overcome the problems created by poorly constructed input. Thus, education that emphasizes clear speech and good writing is a fundamental step.

7.4 COMMUNICATING THE WRITTEN WORD

Documenting is an especially onerous chore for most engineers. Button and Sharrock identified many reasons for this. First, documentation is perceived to be a clerical task rather than a proper engineering or design problem-solving task. It is seen as superfluous and not an integral part of their work. Finally, it is regarded as not being integral to the completion of the task, and therefore it can be postponed or never done at all (Button and Sharrock, 1996).

Engineers are responsible for multiple outputs across many audiences, all of which may require different skills and levels of expertise. Although the amount of communication output varies by the branch of engineering, written communication includes short documents that must convey important information concisely (including memos, instructions, e-mails, meeting minutes, and letters) and more lengthy, formal documents ("prestige documents"), including reports, documentation, proposals, and specifications (Kreth, 2000).

A survey of design engineers in the United Kingdom revealed that they recorded their decisions and design ideas in a variety of media, including diaries, memos, reports, logbooks, files, drawings, and on the computer (Court, Culley, and McMahon, 1994b). Australian manufacturing engineers communicate through several output channels, including: letters, faxes, memos; technical reports; documents; management reports; and proposals (McGregor, 2000). These documents were "seldom the result of a single individual's ideas, knowledge or text. Instead, they are complex amalgams of the work of multiple authors formed according to corporate standards" (McGregor, 2000).

Pinelli, et al. (1988) summarized some aspects of the importance of writing. They reported on a survey by Davis to determine the importance of technical communications to "successful" engineers. Approximately 96% (134 respondents) indicated that the writing they did was either very important (51%) or critically important (45%) in their position, while none of the respondents indicated that their writing was unimportant. Eighty-nine percent of the respondents stated that the "ability to write is usually an important or a critical consideration when a subordinate is considered for advancement."

Spretnak (1982) conducted a survey in 1980, entitled "Technical Communication and the Professional Engineer," that was mailed to 1,000 engineering alumni of the University of California, Berkeley, which asked the question, "Do you have any general comments about the importance or relative unimportance of writing and speaking skills in engineering careers?" None of the respondents indicated that writing and speaking skills were unimportant. Excerpts from the responses to Spretnak's open-ended question include:

- Technical communications is the key to success for every engineer.
- Writing is the most important skill an engineer can possess.
- Writing and speaking should receive the same attention as technical training.

Seventy-three percent reported that good writing skills aided their advancement, 95% said they would consider writing ability in deciding whether to hire or promote an engineer, and 42% of the total respondents said that they would weigh writing and presentation skills "greatly."

The importance of writing to engineering and science students is echoed by Davis (as quoted from Davis, 1975), who states:

> The single, greatest complaint our students make when polled about their undergraduate preparation consists of questions of the form: "Why didn't you teach us how to write?" They have found, much to their amazement, that one of their main jobs in the "real" world is writing, and that they are woefully unprepared to fulfill that part of their duties.

Davis reported that respondents to his study spent approximately 25% of their time in technical writing and approximately 30% of their time working with technical writing of others. Approximately 63% of the respondents reported that as their responsibilities increased, so too did the time they spent writing, and 94% of the respondents explained that they spent more time working with written material as their responsibilities increased. According to Davis, "As their responsibilities increased, respondents spent less of their time developing actual details of specific jobs and more time considering the work of others, making decisions from it, and inaugurating and carrying out appropriate action."

Some technical writers and engineers offer ways to improve poor writing by engineers in the workplace. Writing to the audience and avoiding jargon are two important ways to improve writing (Spencer and Floyd, 1995). Blake (1998) conducted technical writing seminars for working engineers to help address these problems. He pointed out that engineers often experienced problems with writing, including poor organization skills and using "hedging" or "weasel" words. Short workshops and practice can overcome these obstacles.

Recognizing that different forms of communication have different requirements is another way to improve effectiveness. Good correspondence can sell projects to upper management and improve relationships with internal and external customers. Concise writing, neatness, and correctness in memos and letters demonstrate professional competence (Vincler and Vincler, 1997). *The MIT Guide to Science and Engineering Communication* (Paradis and Zimmerman, 1997) covers basic writing and organizational techniques, plus it teaches engineers the best practices for creating a variety of information outputs. There are separate chapters for graphics, meetings, memos, letters, e-mail, proposals, progress reports, technical reports, journal articles, electronic texts, oral presentations, and job-search documents. *A Guide to Writing as an Engineer* (Beer and McMurrey, 1997) also addresses the basics of writing well within the specific context of engineering.

Placing a technical writer on a project team at the outset of a project is recommended by some to help improve product documentation (LeVie, 1997). LeVie cautions technical writers to empathize with engineers' disdain of writing, not to pass judgment on their writing, and explicitly state requirements ahead of time. In teams of five to eight, a technical communicator as project leader can improve misunderstandings, conflict, or crossed purposes (Robinson, 1997).

Management can influence the quality and timeliness of documentation by emphasizing its importance. Button and Sharrock recommend "morally" upgrading "documentation by making it a responsibility to current colleagues, future colleagues, the professors, the population of users, and to the employing organization. The engineer may not like documenting but should recognize that it is important and that it should be recategorised amongst the responsibilities of the profession so that he/she is morally obligated to document." (Button and Sharrock, 1996). Tips for technical writers plus workshops, guidebooks, and articles on how to improve engineers' writing and oral presentations are all attempts to correct existing deficiencies. Educators in schools of engineering around the world are devising courses or curricula that will tackle the problem before new engineers get into the workplace. Chapter 8 delves into the role of education in addressing communication skills.

8

ENGINEERING EDUCATION AND COMMUNICATION SKILLS

8.1 INTRODUCTION

Throughout the 1990s, schools of engineering grappled with how best to incorporate communication courses into their curriculum. Faculty, students, and practitioners widely recognized that both written and oral communication skills were critical for practicing engineers and engineering education was not satisfying those needs. As summarized by Carlson (1999), the "changing realities of the workplace and growing awareness of language in the learning process place added emphasis on Technical Communication in the modern engineering curriculum." During the 1990s, engineering educators introduced many changes, including a renewed emphasis on client-centered design projects, moving toward integrated curricula (Williams, 2000), and focusing on improving communication courses and skills.

Though engineers spend much of their time writing and speaking, they have not done it very well. The decade of the 1990s was a time when engineers' communication skills became the subject of intense scrutiny. Technology has increased the need for better communication skills because engineers now communicate through a growing array of ways to an increasing range of audiences. The accreditation criteria established by the Accreditation Board for Engineering and Technology's Engineering Criteria

Communication Patterns of Engineers. By Carol Tenopir and Donald W. King
ISBN 0-471-48492-X © 2004 Institute of Electrical and Electronics Engineers

(ABET EC2000) specifically mention the importance of communication skills and have driven the move to improve engineering curricula in North America (Williams, 2002). The importance of improving how engineers write and speak is not limited to the United States and Canada, however. Experts from around the world are stressing the need to find ways to help engineers improve their communication skills. The development of new communications technologies, an increasingly global marketplace, and an increased emphasis on teamwork all illustrate the need to improve the written and oral communication skills of engineers in industry, government, and academia.

8.2   IMPROVING ENGINEERS' COMMUNICATION SKILLS

Engineers need expertise in a variety of communication skills in today's workplace. As described earlier in this book, engineers need to know how to write in a variety of types of materials, including technical reports, specifications, patent disclosures, feasibility studies, memos, and e-mail (Schillaci, 1996). Oral communication skills are required for face-to-face communication, telephone, and formal presentations. Most agree that engineering education has not adequately prepared engineers for their communication responsibilities.

Vest, Long, and Anderson (1996) undertook a national survey of electrical engineers to determine whether they felt their education had adequately prepared them in regard to communication skills. Surveys were sent to student electrical engineering members of IEEE in the United States, specifically those who joined IEEE in 1985 or later, restricting the replies to relatively recent graduates. The mean graduation date of the respondents was 1986. Respondents reported that their engineering programs "rarely required them to demonstrate skills in public speaking, presentation, or interpersonal communication" and their preparation in these areas was "poor." Skills in group communication, general writing, and technical writing were "sometimes" required and their preparation in these areas was "adequate." Approximately 69% of the respondents took a composition course outside the engineering department, about one-half had taken technical writing, and about one-third took public speaking. Few enrolled in business writing, interpersonal communication, or group communications courses. More recent graduates reported increased emphasis on engineering courses, group communication skills, public speaking, and senior projects. Over half completed a senior design project, with half of these working alone rather than in groups and over half (59%) presented their project orally to the class.

The electrical engineers who responded to this survey recommended that engineering students learn communication skills within engineering courses and through courses outside schools of engineering. The skills they most frequently recommended were technical writing, presentation skills, and public speaking. About half recommended courses in group communication and "virtually all of the engineers advocate including either formal or informal teamwork experience in the engineering curriculum."

Colorado State University began a five-year project to improve communication skills of electrical engineering students. The university surveyed students, recent graduates, and faculty. The survey revealed that students generally believed they had good writing skills and were highly skilled in using personal computers. Faculty concerns about students' communication skills focused mostly on graduate students, however. Recent graduates report "their primary communication modes are face-to-face discussion and electronic mail" (Zimmerman et al., 1993).

Surveys of faculty and students at Michigan State University found that both faculty and students felt engineering students had trouble communicating. Particular areas of difficulty included grammar, punctuation, and spelling. Other concerns included "lack of organizational skills, unclear expression of ideas, poor verbal skills, difficulty with writing introductions and conclusions, and weak logic." These surveys and earlier interviews with recent graduates have implications for improving engineering curricula. Engineering programs should require courses that include communication skills at least in technical writing and public presentations, and grading should "strongly reflect the communication competence of the student."

Other experts and studies recommend additional ways to improve engineering curricula. Engineering students should also broaden their cultural literacy awareness and sharpen their ethical consciousness (Batts, 1995); technology, such as computer-assisted class notes, self-paced learning tools, and simulated experimentation, should be used in engineering courses to improve communication skills (Chang, McCuen, and Sircar, 1995; Hawthorne, 1999); and there should be inclusion of more written assignments with better feedback, as well as pairing of graduate and undergraduate students to focus on improving communication skills (Gunn, 1995, 1998).

A cooperative program between Arizona State University and Temple University was initiated so students could learn how to work with diverse colleagues from a distance in long-distance collaborative workteams (Barchilon and Baren, 1998). Although the experience was not entirely successful for all the students, seniors from each university worked together to simulate the workplace reality of geographically distributed workteams.

In the spring of 2001, Cornell University and Syracuse University School of Information Studies were jointly awarded a $2.5 million research grant by NASA in order to develop a state-of-the-art virtual learning environment for engineers and engineering students. The goal of this project is to use cutting-edge information technologies to create educational and training situations that improve the abilities of engineers to collaborate on complex, multidisciplinary projects, despite being geographically dispersed.

These efforts reached a peak in 2000 and 2001 in anticipation of the new Accreditation Board for Engineering and Technology's Engineering Criteria 2000 (ABET EC2000) standards that went into effect in the fall of 2001. (See http://www.abet.org.) Six of the 11 criteria of the new standards are described by Williams as "soft skills" rather than technical abilities. She believes "this shift in emphasis" will "have a significant impact on technical communication programs and pedagogy" (Williams, 2000). Learning outcomes are emphasized and evaluation criteria are more flexible, allowing more variation and experimentation in engineering programs (but also creating more confusion among some).

Several ABET EC2000 criteria stress communication-related outcomes, rather than technical engineering skills, including: "an ability to function on multi-disciplinary teams"; "an understanding of professional and ethical responsibility"; "an ability to communicate effectively"; "the broad education necessary to understand the impact of engineering solutions in global and societal context"; "a recognition of the need for, and an ability to engage in life-long learning"; and "a knowledge of contemporary issues." All schools in North America are reevaluating their curricula to meet the ABET expectations of an interdisciplinary curriculum that will be "responsive to the needs of industry and provide an effective background for technical professionals who must solve complex problems in the global workplace" (Williams, 2000).

Later Williams developed five principles for engineering portfolio development as follows:

1. *Defining* engineering communication (or any other learning objective)
2. *Identifying* appropriate skills and mapping them in the curriculum where they are currently (or should be) developed
3. *Correlating* portfolio learning objectives to course and program objectives
4. *Facilitating* opportunities for students to reflect on their learning
5. *Assessing* student learning so that students, faculty, and programs can benefit and improve

The article addressed these five principles to offer guidance to engineering faculty (Williams, 2002).

There have been many approaches to improving communication skills of engineering students over the years. Engineering students have been the beneficiaries of university-wide approaches focused on assisting all students. These approaches have included: Writing Across the Curriculum movements, writing centers to help individual students, required freshman composition courses, and writing intensive courses (Gunn, 1995; Williams, 2000, 2002). The July 1999 issue of *Language and Learning Across Disciplines* was a special issue devoted to "Communications Across the Engineering Curriculum," and the *Journal of Engineering Education* continuously describes efforts to improve engineering education (Randolph, 2000).

Many colleges and universities have described their innovative approaches to enhancing the communications components of their curriculum, but there is no consensus as to the best approach. Three main approaches are used: (1) requiring separate communications courses taught by faculties of technical communication or English; (2) integrating communication skills into engineering courses; (3) relying on internships or projects that teach writing.

The first approach, that of separating communications courses, is now thought ineffectual or, at best, antiquated, though it was the favored approach from the late 1940s through the 1980s. As the discipline of technical writing developed, most now believe separation of communication skills removes engineering context and decreases students' motivation to learn. Academic trends now "increasingly place technical writing instruction within science and engineering classrooms in an attempt to reunite knowledge-making and communication" (Longo, 1997). Engineering faculty may not be expert communicators, however, so technical writing experts fear faculty may teach communication skills incorrectly. Also, engineers worry about their capability to teach communication in addition to technical material. Williams recommends overcoming the "silo mentality" that separates communication and engineering. Technical communicators working with engineering faculty may be the best combination, providing students with multiple options.

The integrated approach may be implemented differently, including: (1) ongoing incorporation of communication skills into engineering classes from the freshman year; (2) offering elective courses focusing on communication skills to be taken either as an undergraduate or graduate student; or, (3) incorporating more communication skills and teamwork into a required senior design project.

Colorado State University added writing to freshman electrical engineering courses by having the required course co-taught by two professors. One taught technical material, and one the writing. Each counted the course as a full course in their teaching load (Mahan, et al., 2000). Writing instruction in the first year of engineering "allows us to establish, early on, our high expectations with respect to report writing for all the engineering courses" (Mahan, et al., 2000). In subsequent courses the instructor's expectation will reinforce the importance of effective communication to engineers.

Engineering schools at many other universities start communications education in the freshman year. The University of California at Santa Barbara, University of Washington, University of Massachusetts, and University of Hawaii are just some of these (Marsh, 1998; Plumb and Scott, 2000; Poli, 1996; Nojima, 1998). At the University of California at Santa Barbara a full-year engineering writing sequence is required for first-year students. The first quarter includes basic writing and research skills in the engineering context. The second quarter builds on these skills and includes skills engineers need to market themselves, including responding to RFPs. The third quarter requires students to publish their work on the Web (Yatchisin, et al., 1998). Other schools emphasize communication skills within a senior capstone course or a senior design project (Dyke and Wojahn, 2000).

Many universities stress the need for an interdisciplinary approach to integrating communication skills with engineering content. The University of Hartford College of Engineering explores communication in a new interdisciplinary course entitled "Engineering Practice" (Nagurney, Keshawarz and Adrezin, 2000). The University of Hawaii revitalized their engineering curriculum in the late 1990s after a survey of faculty, alumni, and students revealed the importance of oral, written, teamwork, and interpersonal communication skills. They feel an interdisciplinary approach is important to meet their expectations. Beginning in the freshman year students are exposed to a curriculum that emphasizes communication skills through repeated practice, cooperative learning strategies, and skills-infused courses.

The U.S. Coast Guard Academy now relies on Writing Across the Curriculum throughout the four-year engineering curriculum. Although this approach adds responsibilities to each engineering faculty member, it has succeeded. Realistic assignments build in complexity throughout the four years and include such information outputs as writing letters, journals, lab reports, essays, sales brochures, and design projects. Seniors must write a publication-quality report as part of their capstone design project (Hiles and Wilczynski, 1995). The University of Texas at San Antonio includes communications emphasis throughout the engineering curricula and selects texts that address composition, technical writing, oral communication skills, visualization, and graphics (Levitt and Howe, 2000).

Hurst and Blicq (1994) recommend a modularly formatted communications course in two intense modules. The first module covers oral and written communication skills at a macro level in the context of engineering; the second takes a micro approach focusing on good writing and oral communication in the workplace.

Providing a context seems to better motivate students and they learn not only how to write and speak better, but how to write and speak like an engineer. Georgia Tech's School of Chemical Engineering offers an undergraduate course that includes oral and written communication in the context of a bioengineering case study that simulates the workplace. Professional engineers are brought in weekly as guest speakers and students gain experience writing press releases, abstracts, and patent disclosures and speaking publicly to different audiences. The course succeeds in its teaching objectives, but as an elective course not many engineering students choose it (Prausnitz and Bradley, 2000).

Second-year electrical engineering students at Rose-Hulman Institute of Technology in Terre Haute, Indiana take a course that exposes them to team projects and improves oral and written communication skills. This is coupled with another cross-disciplinary course in Technical Communication that is taught on the Web in five modules. Each module highlights specific information-processing skills, including: document design, interpreting/presenting data, integrating text and graphics, audience analysis and persuasion, and information and audience engagement (Berry and Carlson, 1999).

Teaching methods influence how well students learn communication skills. A survey of senior and graduate civil engineering students at Lamar University found that students do not support the traditional lecture method of instruction, but prefer a combination lecture/discussion/problem-solving method, which might include working in teams and class discussion instead.

The cooperative (co-op) approach to learning communication skills is recommended by several educators (Lee, 1994; Kreth, 2000). A co-op experience is typically taken during the senior year and involves working with an engineering company in an internship arrangement. Emphasizing oral and written communication in the co-op provides a context and real-world experience, although basic communication skills must be mastered before the co-op experience.

Kreth (2000) recommends the cooperative internship approach combined with formal coursework. A survey of co-op graduates revealed they gained experience in writing reports, memos, instructions, e-mail, letters, specifications, and proposals during the internship. Learning to write like an engineer during a co-op requires feedback from supervisors, writing in engineering courses, feedback from co-workers, and viewing sample documents. Only one-quarter of the respondents to the survey had taken a separate technical writing course and most said such courses are not helpful. Kreth concluded that engineering students benefit more from discipline-specific writing taught during the co-op experience than from general writing courses.

A client-centered senior capstone experience is now required in many colleges of engineering. At New Mexico State University this course was made multidisciplinary in 2000, pairing mechanical and industrial engineering students with technical communication students to work in a team environment. Although not entirely successful yet, the more closely the technical communicators worked with the engineers from the beginning, the more they were valued in the team (Dyke and Wojahn, 2000).

Senior capstone courses in most universities include engineering students working independently or in teams with industry clients. In a survey of recent electrical engineering graduates, over half completed a senior design project, but only half of these worked in project teams. Slightly more than half (59%) present their project orally to the class (Vest, Long, and Anderson, 1996).

Improving engineering education is not just a concern in North America where ABET accreditation drives curriculum development. Engineering projects and companies are increasingly global and need a staff that can work in other countries or with team members in other countries. Dlaska (1999) and Vaughan and Shipway (1995) recommend that departments of engineering and science in the United Kingdom work toward closer integration of engineering content with foreign language courses and cultural awareness to better prepare engineers to work abroad.

Australian engineering students and practitioners also need to improve their communication skills. The Institution of Engineers in Australia identified core communication competencies needed by all engineers. These are similar to, but more specific than, the ABET EC2000 communication standards. Australian engineers should be able to: (1) communicate effectively in the English language; (2) present, report on, and advocate engineering ideas; and (3) prepare, comprehend, and communicate engineering documents. Specifics under each core competency include good oral and written skills, the ability to interpret information and ideas, and skills in presentation, lectures, reports, and so forth (McGregor, 2000).

Integrating communication into engineering courses is used and highly recommended by Nanyang Technological University in Singapore (Collins, Li, and Cheung, 2000). They believe that separate courses are ineffective because the academic training and experience of writing instructors "is far removed from that of engineering," making it difficult for them to develop assignments that are relevant to professional engineering. Nanyang's "task-based approach" relies on integrating engineering content and context with communication skills. Assignments that are relevant to the engineering workplace increase the motivation of students to learn.

As the engineering community becomes increasingly global and communications-intensive, schools like the University of Technology in Sydney, Australia will stress communication throughout their engineering curriculum, emphasizing multicultural, collaborative models. Communications modules are being developed that can be used throughout the engineering program and will include team writing, debating, negotiating, teamwork, writing for publication, writing research reports, and reading and interpreting trade literature. The modules are being developed by the Communications Consulting Program for the faculty of engineering, where communications team members serve as guests or consultants to the engineering experts. Each module would include ways to integrate it into coursework with suggestions for relevant research projects and with theoretical background papers for the engineering faculty (McGregor, 2000).

8.3 IMPROVING THE USE OF COMMUNICATION CHANNELS AND SOURCES

Many skills in written and oral communication are age-old skills, but today's engineering students need to incorporate information technologies to bring their communication skills up-to-date. Civil engineering educators and practitioners met to discuss ways to improve civil engineering education in the United States. They concluded that "today's basic workplace skills are totally different from yesterday's" because "technology has changed both the face of the workplace and the level and types of skills graduates will need when they enter it" (Bakos, 1997). They recommended that the Web be used more in classes and conventional teaching/learning be merged with new technologies, because "communication skills can no longer be identified simply as verbal or written." Engineering students also need to learn to work in teams and learn Web communication skills (Bakos, 1997). Computer-engineering undergraduates in Singapore tended to prefer printed materials and, surprisingly for this discipline, the use of databases and electronic journals was quite low (Majid and Tan, 2002).

Whitmire, in a recent study at the University of Wisonsin at Madison using the Biglan model, observed that engineering undergraduate students relied less than students in other disciplines on library sources of information (Whitmire, 2002). Curl suggests reexamining a model developed by Subramanyam to improve information literacy instruction for engineering students, particularly their understanding of the structure of information (Curl, 2001).

An innovative experiment at the University of Michigan paired library and information graduate students with first-year engineering students. The LIS students served as mentors to the engineering students and taught them about relevant information sources. The experience revealed that "information seekers are frequently unwilling to invest as much time as may be necessary to solve their information needs" and "information seekers may not be so enthusiastic about conducting searches as are information professionals" (Holland and Powell, 1996). Another experimental tactic at Michigan was an Information Resources course taught by librarians for senior engineering students (Holland and Powell, 1995).

Preferences for certain types of materials start while engineers are still in college. Senior engineering students at the University of Michigan prefer their own personal libraries and word of mouth, although a majority also read professional literature and technical materials. Personal knowledge and the knowledge of the members of their working group are highly regarded information resources, as are competitors and informal contacts (Holland and Powell, 1995).

An information resources course was offered as an elective to senior engineers at the University of Michigan, providing an opportunity for Holland and Powell to compare the sources used and valued by students who took the course versus those who did not. Those who took the course subsequently spent 50% more time searching for information and reading than those who did not: 22 hours per month looking for course-takers versus 12 hours for non-takers and 33 hours per month reading by course-takers versus 23 hours for non-takers. Preferred information sources were similar for both groups, but those who took the course rated company librarians more highly than non-takers and showed more interest in nearby college and public libraries. They were not, however, more likely to consult a librarian. In addition, they knew of a broader range of professional technical literature and were more likely to use external information sources (Holland and Powell, 1995).

Poor organization of engineering literature may also contribute to the problems that engineers have using libraries. According to Holland, engineering materials are "profuse, diverse, and sometimes obscure." In some engineering specialties, such as computer architecture and electronics, the half-life of the literature is relatively short, leading Holland to conclude that "as a result, current information is valued, but substantial backfiles in most subspecialties can be a space-wasting dust catcher." Librarians at Texas A&M engineering collection report that it is difficult to locate technical information from engineering organizations due to poor indexing or cataloging and inconsistent naming conventions; papers may be called publications, preprints, reprints, transactions, proceedings, or articles. Still, professional societies are trusted information sources and engineers use them with confidence that technical information from societies will help serve their information needs and help to keep them up-to-date (Holland and Powell, 1995).

Holland and Powell's study shows that the time spent seeking and using information and the range of information sources valued can be positively affected by instruction. They recommend that since engineers use computers and want desktop access, academic librarians should provide engineering students with training and access to many electronic resources. They suggest this because engineering students "show little enthusiasm for approaching a librarian for information," even after taking an information resources class, but "their strong desire to solve problems for themselves carries over to their interest in learning electronic information access" (Holland and Powell, 1995). They also suggest that librarians in academia and corporations should collaborate "to develop a continuum of service for newly graduated professionals" to encourage lifelong information seeking.

Providing information literacy/bibliographic instruction courses specifically for engineering students is another approach recommended by some academic librarians. At UCLA, senior engineering students who took such a course were compared to those who did not. Although no difference was found between the two groups, some of the problems identified in both groups may help improve such instruction in the future. Major problems that hinder information seeking by engineering students include:

- Lack of awareness of the overall information space and sources
- Lack of understanding of some of the basic concepts and tools
- Lack of understanding of the driving forces in the processes of research and publication, such as the publication life cycle, formal versus informal channels of publication, primary sources, secondary access sources, periodical literature, and so on
- Failure to realize which access mechanisms lead to which formats and that formats differ in timeliness and quality (Ercegovac, 1999)

9

THE ENGINEERING SCHOLARLY JOURNAL CHANNEL

9.1 INTRODUCTION

In this chapter we show that engineers spend a smaller proportion of work time reading and read less than other professional groups. Surprisingly, however, engineers rate the importance of this activity to their job as very high. Engineers in academic settings read each article more quickly, and tend to subscribe to more publications than engineers in a nonacademic setting. Information sources other than articles, particularly internal reports and oral communications, are more important for engineers than scientists or medical professionals. Although they read less, engineers spend large amounts of time writing and communicating orally. Recent studies reinforce trends observed for decades, although a growing percentage of reading is now of electronic sources, including e-mail. Between 1995 and 2001, scholarly journals have increased in size.

In this chapter we focus on a particular communications channel, the engineering scholarly journal. The reason for this focus is the extensive use, usefulness, and value of scholarly journals within the scientific and medical communities. While the engineering profession relies less on this channel than the scientific and academic communities (see Chapter 11), journals are ex-

Communication Patterns of Engineers. By Carol Tenopir and Donald W. King   **113**
ISBN 0-471-48492-X © 2004 Institute of Electrical and Electronics Engineers

tremely useful and valuable to the engineering readers. We briefly discuss trends in engineering and science journal publishing at three points in time (1977, 1995, 2001). We also examine engineering authorship and reading of scholarly journal articles and compare these activities with those of scientists. In Chapter 10, we present recent observations of information seeking and reading, and evidence of changes in information-seeking and reading patterns due to electronic publishing.

9.2 ENGINEERING AND SCIENCE JOURNAL CHARACTERISTICS: 1977, 1995, AND 2001

We began collecting data on journal characteristics in 1960 (King et al., 1981) and updated the data in 1995 and 2001, to track how journals are changing, including electronic versions of print. A sample of scientific scholarly journals was examined according to several parameters: annual average number of manuscripts submitted, articles per title, issues, article pages, non-article pages, and number of subscribers. Nine fields of science corresponding to a 1970s NSF distinction were used (physical sciences, mathematics and statistics, computer sciences, environmental sciences, engineering, life sciences, psychology, social sciences, and general science) and a sample of print journals was selected from each of these fields. Information collected for each title included subscription price, number of pages and articles, and number of pages devoted to article and non-article content such as letters to the editor, book or product reviews, special announcements, indexes, tables of contents, or author information.

In 2001, 10 titles were randomly selected from the original 1977 modified title list in each field where possible. Titles without current issues in the library were not surveyed. The overall sample size in 2001 totaled 81 titles. The same physical information was collected for each title in 2001 as in 1995. Two issues from the previous 12 months were randomly selected from each title, and two articles from each issue were randomly selected, with the number of authors, number of citations, and number of pages of each article recorded. *Ulrich's Guide to Periodicals* provided circulation information; *Fulltext Sources Online* provided information on digital versions of the titles. Table 9.1 presents the yearly total averages for journal characteristics.

Table 9.1   Physical Characteristics of U.S. Scientific Journals: 1977, 1995, and 2001

	1977	1995	2001	Increase 1995–2001 (%)
Articles per title	86	123	154	25.2
Issues	6.6	8.3	10.8	30.1
Article pages	644	1,439	1,910	32.7
Non-article pages	188	289	305	5.5
Total pages	832	1,723	2,215	28.6

Source: King, et al., 1981 (1977); Tenopir & King, 2000 (1995).

Table 9.1 shows that from 1977 to 1995 to 2001, the physical characteristics of print journals have increased in every category. The number of articles per title shows an increase with 25% more articles appearing on average from 1995 to 2001. Total pages, however, increased slightly more (29%). There is an overall trend toward price increases due in part to an increase in the size of journals, thereby increasing costs. When these physical characteristics are broken down by the field of science that each title covers, a more complex picture emerges. In Table 9.2, the characteristics of scholarly journals are broken down by the field of science that they represent.

Table 9.2   Average Physical Characteristics of Journals per Title by Field of Science: 2001

Field of Science	Issues/ Year	Subscription Price (Institutions)	Articles/ Year	Article Pages/ Year	Non-article Pages/ Year	Total Pages/ Year
Engineering	11.6	$889	281	2625	355	2980
Physical sciences	20.2	$1,239	653	4373	338	4711
Mathematics and statistics	9.1	$913	107	1903	75	1978
Computer sciences	9.1	$636	57	994	261	1255
Environmental sciences	8.5	$266	99	1375	129	1504
Life sciences	13.7	$252	201	1672	665	2338
Psychology	5.8	$244	68	815	77	892
Social science	3.6	$70	22	480	148	628
General science	16.4	$420	387	2071	1926	3997

Note that engineering journals are the third most expensive type of journal on average ($889), behind mathematics and physical sciences journals. Engineering journals also rank third in number of articles per year (281) and second in total pages per year (2,625).

In a comparison of engineering and all other sciences over the years we find a consistency across the years, as shown in Table 9.3.

The quick growth of the electronic journal can also be illustrated with the samples from the last two surveys. The January 1995 and January 2001 issues of *Fulltext Sources Online* were examined for digital full-text versions of the journals. In 1995, the number of digitized versions available for these journals averaged 0.8; in 2001, the same journals had an average of 4.2 digital versions available from several online aggregators. In 1995, there were no journals in the sample directly available electronically from their publishers. In 2001, from the same sample, 13 (16%) of the journals were available. This parallels the meteoric rise of electronic peer-reviewed scholarly journals and electronic full-text sources, as seen in Table 9.4. According to the *ARL Directory of Scholarly Electronic Journals and Academic Discussion Lists,* only 139 electronic peer-reviewed journals existed in 1995 compared to 3,915 in 2000, an increase in excess of 2,700% (see www.arl.org). Total numbers of printed magazines, journals, and newspaper titles now available in electronic form also grew, although not nearly as precipitously. Total titles grew from 5,646

Table 9.3   Comparison of Journal Characteristics for Engineering and All Other Sciences: 1977, 1995, and 2001

| | Year | | | | | |
| | 1977 | | 1995 | | 2001 | |
Characteristics	E	S	E	S	E	S*
Articles per title	99	84	163	117	281	136
Issues	6.9	6.5	9.0	8.2	9.9	11.0
Article pages	576	652	1,830	1,385	2,625	1,784
Total pages	790	836	2,039	1,679	2,981	2,080
Pages per article	5.8	7.8	11.2	11.8	9.3	13.1

*E = Engineering, S = Science.
Source: King, et al. (1981); Tenopir and King (2000).

Table 9.4 Digital Versions and Digital Availability of Journals by Field of Science: 1995 and 2001

Field of Science	Percentage of Journals with Digital Versions		Percentage of Digital Versions Directly Available from Publisher	
	1995	2001	1995	2001
Engineering	30.0	50.0	0.0	30.0
Physical sciences	0.0	70.0	0.0	20.0
Mathematics and statistics	11.1	77.8	0.0	0.0
Computer sciences	0.0	50.0	0.0	20.0
Environmental sciences	0.0	30.0	0.0	0.0
Life sciences	42.9	100.0	0.0	42.9
Psychology	20.0	50.0	0.0	0.0
Social science	25.0	62.5	0.0	0.0
General science	57.1	81.4	0.0	42.9

Source: Fulltext Sources Online (2001).

sources in 1995 to 15,388 in 2001, a jump of 172% (*Fulltext Sources Online*).

From the data above, the growth of electronic engineering journals seems to lag behind other scientific journals. Only 50% of engineering journals have digital versions as of 2001. However, of the engineering journals available on the Internet, almost one-third (30%) are available directly from the publisher, in contrast to other fields of science, which are generally less likely to be offered in this manner.

9.3 ENGINEERS' AND SCIENTISTS' AUTHORSHIP AND READING OF SCHOLARLY JOURNALS

A national survey conducted in 1977 with engineers and other scientists revealed that only 1.9% of engineers authored an article in the past year, compared with 10.5% of scientists (King et al., 1981). The number of articles authored was 0.023 per engineer and 0.190 per scientist. Thus, engineers clearly did not publish as much as scientists. This holds true today and the size of the journal literature reflects this disparity. In 2001, there were 2,900 engineers per scholarly journal in their field; whereas in science there were only 350 scientists per journal title.

In 1977, most engineers (88%) read scholarly journals, but they averaged only 80 article readings per engineer annually, compared with 128 readings for scientists. Engineers averaged only about 60 hours per year reading these articles, compared with nearly 100 hours for scientists. As described earlier in this chapter, engineers may rely less on formal written sources of information than scientists, but they still spend considerable time reading a variety of written sources. Journal articles are one of the most valuable of these sources. Tenopir et al. (2001) found that engineers at the Oak Ridge National Laboratory (a Department of Energy–contracted laboratory) read an average of 98 journal articles each year. As shown in Table 9.5, there were differences in reading patterns between the fields of science represented among respondents. Engineers read the fewest articles per year compared to physicists and chemists, and also spent the least amount of time reading each year on average. However, the time spent reading per article is nearly twice as high for engineers than it is for physicists and chemists.

Physicists and chemists at ORNL read many more journal articles than the engineers (204 articles each year and 276 articles each year respectively). Engineers spend more time reading each journal article. At ORNL we found that engineers spend an average of 54 minutes per article, while chemists and physicists spend 43 and 45 minutes respectively. Annually, engineers spend an average of about 132 hours reading journal articles, while chemists spend 198 hours per year and physicists spend an average of 153 hours per year. These differences have been consistent in studies over time (Meadows, 1974; Pinelli et al., 1989).

Engineers who published journal articles, reports, or books used more journal articles than other engineers and cited the most relevant of these in their work. Citing rates do not equal rates of reading, however, since many things are read for back-

Table 9.5   Reading Patterns by Scientific Field: ORNL 2000

	Articles per Year	Time Spent Reading per Year (hours)	Time Spent Reading per Article (min.)
Engineers	72	97	81
Physicists	204	153	45
Chemists	276	198	43

ground information or other purposes. Some articles are read and deemed irrelevant. Highly relevant articles may be read more than once (Tenopir and King, 2000a). Georgia Institute of Technology library used citing rates to measure whether their serials collection met faculty research needs. The library serves mostly engineering majors and faculty (63% of graduates in 1990/91 majored in engineering) so most of the serials collection is engineering journals. The librarian found that the collection met faculty needs, because 92% of the periodicals cited were in the library's collection. It could be argued that, given the preference of engineers for ease of use, they tend to cite the readily available journals in the library collections. The authors acknowledge that there may be some validity to this, but point out that faculty also cited gray literature not in the collection (Dykeman, 1994).

A direct relationship between the price of the information and a scientist's or engineer's willingness (or in some cases, ability) to obtain it seems to exist. As serials prices go up, the number of personal subscriptions decreases and scientists rely more on library subscriptions and separate copies through their library or through the Web or other sources (Tenopir and King, 2000a; Tenopir et al., 2001). The number of personal subscriptions of university scientists declined from 4.21 subscriptions per scientist in 1977 to 3.86 in recent years and the proportion of readings from personal subscriptions dropped from 60% to 36% in those years. An even greater difference was observed for nonuniversity scientists. Subscriptions went from 6.20 per scientist in 1977 to 2.44 in the late 1990s, while proportion of readings from them dropped from 72% to 24%. In both instances most of the readings were replaced by readings from library collections (Tenopir and King, 2000a). However, in Chapter 10, we show that the availability of electronic journals has changed information-seeking and reading patterns.

Our studies repeatedly show that scientists who have received awards or are high achievers read more (Tenopir and King, 2000a; King and Montgomery, 2002). This may also trickle down to young engineers who are mentored by engineers who are gatekeepers. Gatekeepers typically read much more than average (Katz and Tushman, 1981; Lee, 1994).

Engineers use all types of published information sources, but they do so in different proportions than scientists and rank each source's value differently. Where scientists tend to rely heavily on peer-reviewed scholarly journals, engineers favor advertising-supported trade magazines (Allen, 1992). King and Tenopir recently reviewed the literature in that discusses the use and readership of engineering and science scholarly journals. They found many relevant studies of journal use since the 1950s. Over time journals have remained an extremely important source of information for scientists. Research results consistently show that scientists[1] rely more on journals than on other sources. Raitt and others also reported that written sources are more important than oral sources for scientists (Flowers, 1965; Mick et al., 1980; Ritchie and Hindle, 1976; Sutton, 1975). The recent SuperJournal Project in the U.K. (1998) found that scientists not only frequently read journal articles (29% daily, 57% weekly), but they also considered journals to be more important to their work (i.e., 84% strongly agree that "journals are important to my work" and another 14% agreed with the statement). Gupta (1981) provides an excellent review of this topic.

Some of the results above apply to scientific literature in general and not just to scholarly journals. However, over the years studies have shown that scientists primarily read journals. For example, in 1956, Shaw (cited by Meadows, 1974) found that 70% of all reading was from journals. In 1968, Gerstberger and Allen compared the use of types of literature: books, professional journals, technical and trade journals, and other publicly accessible written material. Engineers mostly used professional journals and found them to have the highest technical quality; however, they were third in overall accessibility and ease of use. Weil (1977, 1980) reports that journals were read most and provided the most benefits. For the years 1984 to 1998, scientists in several surveys reported their amount of reading of different materials; scholarly journals were always read far more frequently than other documents (Tenopir and King, 2000).

Conversely, engineers and technologists reported that journals were a much less important source of information than interpersonal communication and technical reports. However, these conclusions depend on the type of engineers' organization. For example, a 1989 survey of aerospace engineers (Pinelli et al., 1991)

---

[1]Note that many such studies incorporate engineering into the generic term "science" and, therefore, some results reflect the combined results of scientists' and engineers' readership.

found that engineers in academia used journal articles much more than conference papers, in-house technical reports, and government technical reports (i.e., 26.6 times in a six-month period, 18.0 times, 9.2 times, and 10.0 times respectively). The use reflected the engineers' ratings of importance ranging from 4.35 (on a scale of 1 to 5) for journal articles to 3.02 for in-house technical reports. Use of journal articles in a six-month period by engineers in government was 15.4 times and by engineers in industry was 10 times. Raitt's 1984 survey revealed that oral sources of information were used more by engineers. Others have reported the same results (Gerstl and Hutton, 1966; Ladendorf, 1970; Marquis and Allen, 1966; Shuchman, 1981). Nearly all studies support the finding of Allen and Cohen that "the average engineer makes little or no use of the scientific and professional engineering literature." Instead engineers rely heavily on internal technical reports and personal contacts, although Allen (1964) found an inverse relationship between performance and use of outside persons (e.g., consultants). Similar conclusions were made by Shilling and Bernard (1964); Auerbach Corporation (1965); Rosenbloom and Wolek (1967); Gerstberger and Allen (1968); Allen et al. (1968); Beardsley (1972); Gerstenfeld and Berger (1977, 1978, 1979); Krikelas (1983); and Pinelli et al. (1989).

Several studies over the years report the number of journals used by scientists. For example, in 1948, Bernal (cited by Meadows, 1974) reports that the average number of journals consulted per week was 5 to 15 according to the group surveyed (9 for scientists and 5 for engineers). Menzel et al. (1960) found that 60% of reading by chemists was from 3 main journals and 25% of reading by zoologists was from 3 journals. Martin in 1962 (cited by Meadows, 1974) indicates that 10 journals accounted for half the reading done by chemists and physicists. Allen and Cohen (1969) indicate that the "stars" in one organization read an average of 8.2 journals in one setting and 4.4 journals in another setting. Others, in both settings, read 3.6 journals on average. Frost and Whitely (1971) found similar results; that is, stars read 6 scientific journals regularly and others 3.1 (median). In 1970, Wolek observed the ranges in the number of publications that are read regularly by researchers and engineers. These average 6 publications for researchers and 4.3 for engineers. Taylor in 1975 found that the number of periodicals read regularly was 18 for technical discipline choices, 29 for testing technical idea choices, and 14 by technical discipline stars. A 1990 survey of 156 researchers in 6 companies found they had reviewed journals, but they shared an average of 8.3 papers and journals with colleagues (Mondschein). In response to the question; "On average, how often do you use electronic journals?" scientists in 1999 answered 8 daily, 30 weekly, 9 monthly, 13 occasionally, and 10 never (Pullinger and Baldwin, 2002). Surveys of scientists in 1977 observed that scientists read at least one article from 13 journals; in a series of surveys in the 1990s that number increased to 18 journals (Tenopir and King, 2000). It appears that scientists are reading a wider range of journals now than in the past.

Below we review studies that provide estimates of the extent to which individual engineers and scientists read scholarly journals. These studies invariably involve surveys of individuals, although the way in which the questions of readership and time are asked varies somewhat. Surprisingly, few of the many surveys of scientists' information-seeking behavior ask about the number of articles read. In 1948, Bernal (cited by Meadows, 1974) estimated that medical researchers read an average of 7.4 papers per week (perhaps 340 to 380 per year) and engineers read 1.5 papers per week (about 70 to 80 per year). A 1977 national survey showed that scientists averaged 105 article readings per scientist (King et al., 1981), while a follow-up survey in 1984 showed about 115 readings, and several surveys in organizations from 1993 to 1998 yielded combined estimates of about 120 readings per scientist (Tenopir and King, 2000). Engineers are found to read less than other scientists. For example, Pinelli et al. (1989) estimated that engineers read an average of 6.7 articles per month (or about 80 readings per year) and the engineers in the surveys above also averaged 80 readings per year (both of these results are about the same as the 1948 Bernal observations). Thus, evidence suggests that the amount of reading by scientists and engineers has not changed much over the years.

Several studies have shown that academicians read more than nonuniversity scientists. In 1969, Meadows and O'Connor revealed that university scientists use journals more than those in government establishments. King et al. (1981) estimated that university scientists read scholarly journals an average of 150 times per year versus 90 times by other scientists. In the period 1993 to 1998 several surveys produced averages of 188 and 106 readings of scholarly journals per university and nonuniversity scientist respectively (Tenopir and King, 2000). However, since there are many more scientists working outside of universities, they account for about 70% of all readings. The Association for Computing Machinery (ACM) provides some confirmation of these results (Denning and Rous, 1994). They say that most ACM "journals are written by experts for other experts, but these experts constitute less than 20% of the readership." Note that most scientific scholarly articles are written by university scientists. The other 80% of the readers of ACM journals are said to be experts from other disciplines or practitioners.

Several studies have investigated the hours or proportion of time spent reading the literature. For example, Halbert and Ackoff (1959) estimated that in 1958 physical scientists spent about 37 hours per month reading. In the 1960s, the Case Institute of Technology Operations Research Group reports 24 hours per month. Hall et al. report 45 hours, Allen (1966a) reports 8.6 to 13.8 hours, and Garvey and Griffith (1963) report 27.7 hours for psychologists and 15.6 to 20.8 hours for other fields of science. Raitt (1984) reports time in increments, which roughly convert to 15 hours per month for background reading by aerospace engineers, and Holland and Powell (1995) indicate that University of Michigan engineering graduates spend about 22.9 hours per month reading. Results, while varied, do indicate an appreciable amount of time spent reading. Hinrichs (1964) reports 10% of time, Mick et al. (1979) report 9.8%, Allen (1966a) reports 7.9% for engineers, and in 1988, 18.2% for scientists (including about 5% involving electronic messages). Again, engineers' time is a scarce resource that is carefully utilized. A decision to spend an appreciable amount of time reading suggests that they place considerable value on the information received. While not all reading involves scholarly journals, studies suggest that amount of time spent reading journals is substantial.

The purposes for which scholarly journals are used have been described in a variety of ways over the years. For example, Allen (1966b) determined the proportion of times information sources were used for various purposes on research projects. The proportion of times the literature is used for various purposes is as follows: expand alternatives (60%), generate alternative approaches (58%), generate critical dimensions (54%), set limits of acceptability (50%), test alternatives against dimensions (27%), and reject alternative approaches (13%). Literature is the most frequently used source for all purposes except the last two. In 1967, Rosenbloom and Wolek found that in central laboratories, professional documents are used for research (48%), design and development (33%), and analysis and testing (48%) and that they are the most frequently used channel for all three purposes. Garvey et al. (1974) revealed that journal articles are used by scientists to: (1) form a basis for instruction of new scientists, (2) acquaint themselves with the accumulated knowledge that exists when embarking on new research or inquiry, (3) facilitate day-to-day scientific work, and (4) advance the research front.

Others also provide data on use of journal literature. Machlup and Leeson (1978) found in their survey of economists that general interest is the most prominent purpose (46%), followed by research (33%), teaching (15%), and coursework and other purposes (6%). More recently, Sabine and Sabine (1986) established that journals in 50 libraries were used for current research (38%), help on the job (25%), writing a paper or speech (13%), general information (10%), and teaching (5%). Berge and Collins (1996) observed that electronic journals were used because of interest in a topic (68%), to help in work (25%), or for researching a topic (14%). In 1998, Shoham showed that purposes vary by the type of scientist (see Table 9.6).

Tenopir and King (2000) make a distinction between university and nonuniversity scientists' purposes for using scholarly journal articles. In 1993 over 50% of readings by University of Tennessee scientists were for current awareness or professional development. Other readings were used to support research (75%), for teaching (41%), to prepare formal publications and formal talks or presentations (32%), and for administration (13%). C.M. Brown also found that scientists at the University of Oklahoma relied on journals more for research than for teaching. On the other hand,

Table 9.6   Purposes of Reading by Engineers, Scientists, and Social Scientists

Purpose of Use	Engineers	Scientists	Social Scientists
Research	78.6%	94.9%	90.1%
Instruction	21.4%	41.8%	72.7%
General updating	71.4%	75.9%	79.1%
Obtain research funds	14.0%	10.1%	14.5%

Source: Shoham (1998).

Griffiths and King (1993) found that scientists in 32 nonuniversity settings (e.g., AT&T Bell Laboratories, National Institutes of Health, Oak Ridge National Laboratory) used journals differently: for current awareness or professional development (30%), background information research (26%), conducting primary research (17%), conducting other R&D activities (11%), and management or other (3%). Their readings for communications-related activities were: consulting or giving advice (4%), writing (7%), and making presentations (3%).

Information in journal articles is found to be important for a number of reasons. Scott (1959) reports that literature is the primary source of creative stimulation for scientists and engineers. The user's perception of the quality of the information is a major factor in both the use of information and the adaptation of innovation (Chakrabarti and Rubenstein, 1976). The literature of a profession was shown to be the single most important source of information in achieving product innovations (Ettlie, 1976). Another aspect of the usefulness of scholarly journals is their importance to scientists. Machlup and Leeson (1978) report that economists found 32% of their readings to be useful or interesting, 56% moderately useful, and 12% not useful.

Tenopir and King (2000) indicate the importance of journal article readings to the purposes mentioned above. University scientists rated importance from not at all important (1) to somewhat important (4) to absolutely essential (7). For readings done for teaching, scientists rated the importance of the information in achieving teaching objectives as 4.83 on average, while importance to research was given an average rating of 5.02. Over a period of a year, the scientists indicated that, of a total of 188 readings, an average of 13 readings per scientist were absolutely essential to their teaching and 23 were absolutely essential to research. Nonuniversity scientists were asked to rate the importance of several resources (e.g., computing equipment/workstations, instrumentation, documents, advice from others, etc.) used to perform various activities. The ratings were from 1 (not at all important) to 5 (absolutely essential). The average ratings of journals for activities performed are as follows: professional development (4.05, highest), primary research (4.03, second highest), other R&D activities (3.87, highest), writing (3.76, highest), consulting/advising (3.60, second highest), and presentations (3.31, third highest).

9.4   CHANGES IN INFORMATION-SEEKING AND READING PATTERNS FOLLOWING ELECTRONIC JOURNALS

In order to provide a before-and-after picture of the effects of electronic journals, we surveyed engineers and scientists at the Oak Ridge National Laboratory, first in 1984 and again in 2000. Between 1984 and 2000, there were differences in the amount of reading, source of articles read, how the articles were identified, and time spent obtaining and reading articles. In all of our surveys we have defined reading as "going beyond the table of contents, title, and abstract to the body of the article." To be current, we stated in the 2000 survey that "articles include those found in journal issues, author Web sites, or separate copies such as preprints, reprints, and other electronic or paper copies." The estimated amount of "readings" includes multiple readings of one article. In fact, about 17% of the 2,000 readings involved articles that had been read prior to the most recent reading. This occurred more often with paper-based articles (22%) than with electronic/digital articles (4%).

In 1984, we estimated that the ORNL scientists and engineers averaged 99 journal article readings per year, and in 2000 the average was 113 article readings—an indication that the amount of reading of articles may be increasing. This phenomenon is consistent with over 13,500 survey responses from engineers and scientists observed from 1977 to 1998 (Tenopir and King, 2000a).

What has changed over time are the sources of the articles, both in proportion and amount of reading (see Table 9.7).

The most striking differences in sources from 1984 to 2000 were the increase in the proportion and amount of readings from

Table 9.7   Proportion and Average Amount of Readings per Person from Various Sources of Articles: ORNL 1984 and 2000

Source	1984		2000	
	Proportion (%)	Amount of Reading	Proportion (%)	Amount of Reading
Personal subscription	37	37	29	33
Library collection	53	52	48	54
Shared dept/unit collection	2	2	3	3
Separate copy	8	8	18	23

separate copies and the shift from personal subscriptions to library collections.

One consequence of this change in behavior was that scientists appear to be reading from a larger number of journals. The 2000 survey showed that respondents read at least one article per year from approximately 23 journals. While we do not have comparable data from the 1984 survey, other surveys of scientists indicate that the number of journals from which a scientist read articles in a year rose from 13 in the late 1970s to 18 in the 1994 to 1998 time period. Some of the changes observed in the range of journal titles and amount of readings from separates were due to an increase in readings of articles identified by online searches (7.5% of readings in 1984 to 13.3% in 2000) or recommended by other persons, such as colleagues (8.6% and 24.0% respectively). Observations from OhioLINK and others confirm that users read from a wider variety of titles when the literature is made available electronically.

Another difference observed from 1984 to 2000 was the proportion of readings from electronic journals and digital databases. There were no readings from these media in 1984, but in 2000 about 35% of the readings were from them. Over one-half of these readings involved browsing electronic subscriptions provided by the ORNL libraries (18% of readings), free author websites (2.7% of readings), or personal electronic subscriptions (1.3% of readings). Nearly all of the browsed electronic journals were published in 2000, but one respondent reported a publication date of 1999. Another 5% of the readings were from electronic library subscriptions but were identified from citations in other publications or from online searches; 5% were from personal electronic subscriptions involving articles identified from citations in other publications or mentioned by other persons; and 4% were from websites with articles mentioned by other persons. These readings were nearly all year-2000 publications, but one reading from a personal electronic subscription was from a 1990 publication.

The proportion of readings found by browsing did not change much over time. In 1984, about 41% of readings were found by browsing personal or current library collections, and 6.5% were found by browsing copies routed by the library (i.e., 48% total browsing). In 2000, a total of about 45% of readings were found by browsing personal print subscriptions, library print subscriptions, department collections, and electronic or digital copies as mentioned above (see Table 9.8).

In 1984, about 13% of readings were identified in printed indexes, but use of printed indexes dropped to zero in the 2000 survey. Readings identified by citations in other publications dropped from 24% in 1984 to about 7% in 2000. The distribution of the age of articles read sheds some light on reading patterns as shown in Table 9.9.

The amount of reading of articles over one year old remained similar for the two time periods; however, in 2000, there appeared to be substantially more new articles read (i.e., 80.2 readings per person in 2000 versus 59.4 in 1984). Nearly all the shift to recently published articles was attributable to reading of electronic or digital articles. Of all the articles read from electronic or digital media, 85% were published in 2000 (8 months into the year), while only 56% of articles read from print subscriptions or copies

Table 9.8   Proportion of Readings Found by Browsing from Four Sources: ORNL 2000

Source	Proportion Found by Browsing (%)
Personal print subscriptions	20
Library electronic	20
Library print subscriptions	4
Department collections	1

Table 9.9   Average Number of Readings of Articles per Person by Age of Article Read: ORNL 1984 and 2000

Age of Article	Readings per Person	
	1984 (%)	2000* (%)
1 year	59.4	80.2
2 years	12.9	10.2
3 years	5.9	4.5
4–5 years	11.9	6.8
6–10 years	4.0	4.5
11–15 years	3.0	2.3
over 15 years	4.0	4.5

*Readings adjusted from 8 months to a year.

were published in 2000. The oldest article read in the 2000 survey was 25 years old.

The fact that the electronic/digital reading tended to be of more recent articles means that fewer of the articles had been read prior to the most recent reading (4% in 2000 vs. 22% in 1984). In reading from both digital and print journals, a high proportion of the readings involved information that was known by the scientist prior to the first reading of the article (44% electronic reading vs. 58% print). In both instances such articles were often found from citations in other publications or after mention of the article by another person.

Time spent identifying, locating, and obtaining the articles changed since 1984 in a way that might not be expected: the time per reading spent browsing or searching for the article and determining where the article was located approximately doubled, according to the 2000 survey. The reported time spent browsing electronic/digital articles was estimated to be 13.3 minutes per reading, but the time spent browsing print copies was half of that time (6.5 minutes). The time spent obtaining or accessing the article was about the same in the two surveys (7 and 6 minutes respectively). When time involving other activities such as locating, displaying, and downloading or printing was added, the time spent totaled 17.7 minutes per electronic/digital reading. This was compared with 8.2 minutes for browsing print copies (including locating and photocopying the articles). About 38% of the electronic/digital readings were read from the screen. These readings tended to be of shorter duration than the downloaded/printed readings (i.e., 20 vs. 62 minutes, respectively). Interestingly, the proportion of print articles photocopied was about 50% compared with 62% of electronic/digital articles downloaded/printed out. The time spent photocopying was about three minutes compared with 4.5 minutes spent downloading/printing. When articles were identified by means other than browsing, the time spent using the two media was about the same (i.e., 22 minutes per reading electronic/digital articles and 19 minutes for print articles).

The principal purposes of the information obtained from the articles read were most frequently primary research (34% of readings), background research (24%), and current awareness or continuing education (22%). These proportions of readings tended to be slightly higher for electronic/digital articles than for print articles. About 16% of the readings were for communications-related purposes such as writing, making presentations, or consulting/advising others. Other purposes, such as administration, accounted for the remaining purposes of reading.

The respondents surveyed in 2000 indicated that they averaged 98 hours per year reading journals (96 hours in 1984). This estimate was based on estimated amounts of reading (99 readings in 1984 and 113 in 2000) and the average time spent per reading (58 minutes and 52 minutes, respectively). Because their time is a scarce resource, this amount of time spent was an indicator of the value of the information gained from reading journal articles. The amount of time spent reading electronic/digital articles was nearly identical to that of paper-based articles (i.e., 52.2 minutes per article versus 51.4 minutes). Thus, this indicator of value was also the same for the two media sources. Other indicators of value of information include the observation that respondents whose most recent reading was from an electronic/digital article tended to be older and publish more articles.

In some scientific fields, preprints of journal articles are an important distribution means. Physics, particularly high-energy physics, is an example. In a 1977 national survey of scientists, it was estimated that scientists received and read 2.1 million preprints (King, McDonald and Roderer, 1981). In 1981, physicists read about 20,000 separate copies of articles from 19 American Institute of Physics journals; 4,500 of them were preprints. Physical science authors distributed an average of 110 preprints per article (King and Roderer, 1978). Several digital preprint services have evolved in recent years, including the Los Alamos National Laboratory arXiv.org e-print archive and the DOE PrePRINT Network. (LANL's arXiv also includes electronic articles other than preprints. DOE's PrePRINT Network is a gateway service to nearly two dozen separate preprint or e-print servers.) We did not specifically include preprint reading in the 1984 survey but rather included it as part of a general category of separate copies of articles. In the 2000 survey we asked respondents about their awareness of these (and other) preprint services, how much reading they did from them, and whether they submitted articles to the services.

About 29% of the ORNL respondents were aware of the LANL archive service, and about three-fourths of those who were aware had read 7.9 preprints per person from the service in the past 12 months. Roughly one-half of physicists were aware of the LANL services, and nearly all of those aware had read preprints from it in the past year. Other fields particularly acquainted with the service included engineering (31% aware) and chemistry (20% aware). Of all the respondents aware of the service, only 14% of them had ever submitted article preprints to arXiv.org, even though those respondents averaged authoring or co-authoring about 8 articles per person in the last two years. About 10% of the articles published by those aware of arXiv.org were submitted to the LANL service. A similar proportion of respondents (25%) were aware of the DOE PrePRINT Network, but fewer of them (53%) actually read preprints mentioned by the service. Those who did so averaged reading six preprints per person in the last year. Most of these readers were physicists or engineers. Other services were mentioned and used by a few of the respondents, including such websites as Physics of Plasmas, IOP, and Nuclear Fusion; ACM; and High Tc Update.

Altogether, the total electronic preprint reading amounted to about 3.6% of all reading. In addition, about 4.5% of readings were from preprints sent to respondents for article review or refereeing. Since about one-half of reading from separate copies of articles involved preprints, the increase in amount of reading from those separate copies may be partially attributable to reading from preprints and corresponding preprint services.

10

ENGINEERS' JOURNAL INFORMATION-SEEKING AND READING PATTERNS IN AN EMERGING ELECTRONIC ERA

10.1 INTRODUCTION

This chapter focuses on the information-seeking and reading patterns of engineers in the emerging electronic journal era. It is clear that electronic journals are having a significant impact on journal communication patterns of scientists (King and Montgomery 2002; Tenopir et al., 2003) and medical professionals (Tenopir, King, and Bush, 2003). From 2000 to 2003 readership surveys were conducted at the Oak Ridge National Laboratory (ORNL), University of Tennessee, Drexel University, and University of Pittsburgh. In order to examine information-seeking and readership patterns of engineers, 98 observations of engineers were extracted from these surveys for the analysis given below. It is emphasized that the engineers at ORNL are not typical of nonuniversity engineers, but do give some indication of the contrast between patterns of engineers located in universities and elsewhere.

In particular we rely on readership surveys that emphasize observations of a "critical incident" of the last reading of a journal article. This method permits one to examine each reading to develop the multiple patterns of how engineers learn about the articles they need (e.g., browsing journals, searching, being told

Communication Patterns of Engineers. By Carol Tenopir and Donald W. King   **133**
ISBN 0-471-48492-X © 2004 Institute of Electrical and Electronics Engineers

about them), where they get these articles (e.g., personal subscriptions, library collections, separate copies), the format or medium of the articles when read (e.g., print, photocopy, electronic—on the screen or printed out), age of the article when read, time required to obtain and read the articles, outcomes from reading the articles, and so on. This approach recognizes that every reading is unique and results in a specific combination of those factors.

The chapter is subdivided into sections on the use, usefulness, and value of journals, where engineers get articles they read, format of the articles, how they learn about the articles, issues dealing with age of the articles read, and factors that influence these aspects of article information seeking and reading. The article use is observed by how many articles engineers report that they read in the past month (projected to an annual amount). A "scholarly" article is defined as one "found in journal issues, author websites, or separate copies such as preprints, reprints and other electronic or paper copies." A reading is defined as "going beyond the table of contents, title, and abstract to the body of the article." With the exception of electronic-related aspects, these definitions have been consistent with over 25,000 readership survey responses dating back to 1974. Usefulness is examined in terms of the purposes for which articles are read and the importance of information in achieving the principal purpose. The "value" of the article is defined in two ways: by the consequences of reading (described above) and by what engineers are "willing to pay" for the information. Engineers pay for information in (1) dollars paid for subscriptions or related services and (2) in their time required to obtain and read articles. The latter tends to be 5 to 10 times higher (when a dollar amount is assigned to time) than payments made.

Engineers get articles from personal subscriptions or library collections in electronic or print format. They can also receive photocopies, reprints, or electronic copies of articles from colleagues, authors, or other persons. Libraries also provide copies through interlibrary loan or document delivery services. In some professional fields, including engineering (Lawal, 2002), electronic preprint archives and author websites are a source of copies of articles. Engineers often learn about the articles they read by browsing personal or library subscriptions (print and electronic) or other digital collections; usually from recently published articles. They can also search for articles using an index and abstract

(A&I) bibliographic database (e.g., Engineering Index, Compendex), web search engine, online journal collection, current awareness service (print or electronic), preprint service, or other e-print service. Finally, engineers also learn about articles through citations in other publications or being informed about them by others.

There are some issues concerning the age of the articles read. For example, there is a question of whether older articles continue to be read with the emergence of electronic journals. This is important because older articles tend to be more useful and valuable than recently published articles. If the older articles continue to be read, it is important to learn where they are obtained, how they are identified, and in what format they are read. There are several factors that affect how engineers learn about the articles read, where they obtain them, and the format used. These and other issues are discussed in the sections below.

10.2   USE, USEFULNESS, AND VALUE OF ARTICLES TO ENGINEERS

Journal use is defined in many ways (King and Tenopir, 2001); reading as defined in Section 10.1 is the measure used below. The amount of reading done by engineers is observed by asking them: "In the past month (30 days), approximately how many scholarly articles have you read?" A reported number of readings is *not* the same as number of articles read, because an article can be read once or many times. For example, an article might be read to keep up with the literature when it is first published, but read again later when an information need arises. In fact, for the "critical incident" of reading the engineers are asked if they "had read this article prior to this particular reading?" As it turns out, about 30% of the readings had been read before, which is typical of science in general. Another aspect of this survey question is that the distribution of readings reported tends to be highly skewed. That is, a few engineers have read a great deal in the last month, say, over 100 readings, but most report fewer than 10 readings (i.e., about 70% of the engineers reported 10 or fewer readings in the last month). In previous readership studies respondents who report extensive reading are sometimes contacted to make sure they meant that the large reported amount was for a month and not a year. They nearly always did in fact read heavily during the month reported.

The average number of readings reported for the last month was 13.5 readings per engineer or about 162 annual readings per engineer. The average readings per university engineer was 186 annual readings and at ORNL the average was 98 annual readings. The estimated averages among the three universities are fairly consistent and reflect what one might expect of reading by university engineers, but ORNL is atypical of nonuniversity average amount of readings. In 1977, a national survey of scientists yielded an estimate of 80 annual readings per engineer (King et al., 1981). There is an indication that the amount of reading per engineer may actually be increasing some since 1977.

At ORNL the scientists averaged 146 annual readings (compared with 98 engineers' annual readings) and at the universities surveyed, scientists averaged 205 annual readings (compared with 186 engineers' annual readings). In the 1977 national survey the scientists averaged 128 annual readings. National estimates for both engineers and scientists reflect smaller average numbers of readings because most engineers and scientists work outside of universities and typically read much less than university members. In fact, if the estimate of 186 annual readings by university engineers held for all university engineers one would expect the nonuniversity engineers' average annual readings to be, perhaps, about 75 readings per engineer.

Value of article information content is measured in two ways: (1) by what engineers are willing to pay for the information in dollars and their time and (2) by the use value represented by consequences of reading. Over the past 25 years, the number of personal subscriptions of scientists has declined from about 5.8 in 1977 to 2.2 subscriptions per scientist currently. However, engineers have never been heavy subscribers. The 1977, national survey yielded an estimate of 1.2 subscriptions per engineer and in the 2000 to 2003 era the average number of subscriptions per nonuniversity engineer at ORNL is estimated to be 1.16 and for university engineers 2.56 (16% of university engineers and 36% of ORNL engineers report that they receive no personal subscriptions). Most of the personal subscriptions are derived from engineer society membership, although a very small proportion of the university personal subscriptions are obtained through grant funding.

The point is that engineers pay a fairly modest amount for the journals they read. On the other hand, they are willing to spend an appreciable amount of their time reading journal articles. The engineers surveyed by us from 2000 to 2003 average spending 88 hours per year reading articles at ORNL and 127 hours at the universities. In many respects, engineers' time is their most scarce resource and they would not be willing to spend this time reading if they did not consider the information obtained to be of corresponding value. Over the years we have observed the dollar value of readers' time spent reading to be 5 to 10 times the amount expended for subscriptions (including a prorated amount for library purchases).

It was also observed that male engineers in the universities appeared to read more than female engineers (190 vs. 160 annual readings per engineer), but while female engineers may read fewer articles, they appear to take more time reading them than male engineers (50 vs. 40 minutes per article read for male engineers). Thus, they annually spend about the same amount of time reading (133 vs. 127 hours per year for male engineers).

Value is also measured in terms of the consequences of reading journals. One indicator of such value is the purpose for which articles are read. For the critical incident of last reading we asked engineers to indicate the principal purpose for which they have used or plan to use the information obtained from the article. As shown in Table 10.1, the principal purpose is primary research for both

Table 10.1   Proportion of Article Readings by University and ORNL Engineers by the Principal Purpose for which the Article Is Read: U.S. 2000–2003

Principal Purpose for Reading	University (%)	ORNL (%)
Primary research	58.7	32.1
Background research	23.8	21.4
Teaching	6.3	*
Writing reports, articles, etc	4.8	*
Presentations	*	7.1
Consulting, advising others	1.6	10.7
Current awareness, keeping up	4.8	25.0
Administration	*	3.6
Total	100.0	99.9

*No observations
Source: Surveys at ORNL, Drexel University, University of Pittsburgh, University of Tennessee (*n* = 91).

university and ORNL engineers. Background research is the second most frequently reported principal purpose of reading by university engineers, with teaching a far third.

However, at ORNL the engineers report primary research far less frequently than the university engineers (32% vs. 59% of readings). While background research is reported about equally, the ORNL engineers read articles far more frequently for current awareness or keeping up with the literature (25% vs. about 5% of readings).

We also asked engineers to indicate purposes for reading other than the principal purpose. In Table 10.2, all of the reported purposes are given (not just the principal ones). When all purposes are considered, teaching, writing, and current awareness become frequent purposes for university engineers, and at ORNL current awareness and continuing education become much more frequent purposes.

In the university surveys, the engineers were also asked: "How important is the information contained in this article to achieving your principal purpose?" The ratings were from 1 (not at all important) to 4 (somewhat important) to 7 (absolutely essential). Overall, about 7% of readings were rated as being absolutely essential, mostly to primary research. The average importance rating for primary research was 5.49 and 4.48 for all other principal purposes. Thus, a substantial proportion of reading is important

Table 10.2 Proportion of Article Readings by University and ORNL Engineers by the Various Purposes for Which the Article Is Read: U.S. 2000–2003

Various Purposes for Reading	University (%)	ORNL (%)
Primary research	77.8	42.9
Background research	58.7	50.0
Teaching	33.3	*
Writing reports, articles, etc.	52.4	21.4
Presentations	19.0	21.4
Consulting, advising others	22.2	21.4
Current awareness, keeping up	47.6	60.7
Continuing education for self	1.6	46.4
Administration	3.2	3.6

*No observations.
Source: Surveys at ORNL, Drexel University, University of Pittsburgh, University of Tennessee ($n = 98$).

to the engineers' research and other endeavors (i.e., about one-third of readings were rated 6 or 7). The surveys also sought to learn: "In what ways did the reading of the article affect the principal purpose?" When the principal purpose was primary research, the most often cited way was that "It improved the result" (80% of these readings), followed by "It inspired new thinking or ideas" (54% of readings). Other ways included "Narrowed, broadened, or changed the focus of research" (39%), "Resolved technological problems" (20%), "Saved time or other resources" (17%), "Resulted in collaboration or joint research" (9%), and "Resulted in faster competition" (7%). Similar reasons were observed for the other principal purposes, but with much less frequency.

One indicator of the use value of articles to engineers is that engineers whose research or other profession-related contribution has been acknowledged through awards or other special recognition tend to read more and spend more time reading than those whose work has not been acknowledged. Engineers surveyed who received such recognition averaged 225 annual readings compared with 126 readings by those not recognized and they spent about 80 more hours reading. This has been a typical observation over the years (Tenopir and King, 2000).

10.3 WHERE ENGINEERS GET THE ARTICLES THEY READ

From the critical incident of last reading, the surveys determined the source of articles read. As shown in Table 10.3, the three principal sources are from personal subscriptions, library collections, and separate copies. The sources used are not too different for university and ORNL engineers, with a higher proportion of readings from library collections by ORNL engineers (50% vs. 43% of readings), and a little lower for the other sources. The subproportions of readings from print and electronic formats are given in brackets for personal subscriptions and library collections. That is, 88% of readings of university personal subscriptions are from print issues. Few readings of personal subscriptions by university engineers are from electronic format, but nearly two-thirds of their library collection readings are from electronic journals. The opposite pattern is observed for ORNL engineers. This pattern shows that the emergence of electronic journals appears not to influence the need by engineers for library access to journals (noting

Table 10.3   Proportion of Article Readings from Various Sources by University and ORNL Engineers: U.S. 2000–2003

Source	Universities (%)	ORNL (%)
Personal subscriptions	**35.7**	**32.1**
Print	[88.0]	[55.6]
Electronic	[12.0]	[44.4]
Library collections	**42.9**	**50.0**
Print	[36.7]	[64.3]
Electronic	[63.3]	[35.7]
Separate copies	**21.4**	**17.9**
Preprint-electronic	[6.7]	[*]
Reprint	[13.3]	[*]
Another person	[46.6]	[60.0]
ILL/Doc. Dev.	[20.0]	[*]
Review copy	[13.3]	[40.0]
Total	100.0	100.0

*No observations.

Source: Surveys at ORNL, Drexel University, University of Pittsburgh, University of Tennessee (*n* = 98).

that with Interlibrary Loan and document delivery, the library access in universities increases to 47%). This phenomenon is undoubtedly necessitated by the low number of personal subscriptions obtained by engineers, requiring them to acquire articles from other sources such as their library.

Separate copies of articles have long been an often-used source of articles. About 16% of readings by engineers were observed to be from separate copies in the 1977 national survey and in the current surveys a similar, but slightly higher, proportion was observed; that is, 21% of university readings and 18% of ORNL readings. The most frequent source of separate copies is from another person such as a colleague, author, and so on.

10.4 FORMAT OF ARTICLES READ

About one-third of the articles read by engineers are from electronic journals or electronic separate copies and the proportion is almost identical for university and ORNL engineers (although from different sources). Of the electronic article readings, most are printed out (76% of electronic article readings), although some continue to be read on the screen (24% of the readings). Engineers also tend to photocopy from print format. In fact, 6.8% of readings from print personal subscriptions are from photocopies and an even greater proportion of reading from library collections are photocopied (45%). Gender is not a factor in the format used by engineers, but age may well be. A proxy for age is the number of years since receiving an engineer's highest degree. At 1 to 10 years since graduating, the engineers tend to read a higher proportion of electronic versions than older engineers: 48% of electronic readings vs. 45% at 11 to 20 years, and less than 30% over 20 years.

10.5 HOW ENGINEERS LEARN ABOUT THE ARTICLES THEY READ

An important information-seeking activity involves identifying and locating needed articles. Generally, engineers learn about the articles they read by browsing journals, or searching for specific information, or other persons told them about the articles, or the articles were found as citations in other publications. Browsing tends to be done on recently published journals to keep current with the literature. The engineers will find some articles that are of immediate use, but often will identify articles that are of interest, but not of current use. In this instance, they will frequently photocopy or print out the article for future reference. Above we mentioned that 30% of the readings involved articles that had been read before and many of the previous readings initiated through browsing. As shown in Table 10.4, about half of readings are identified through browsing. Note that the "critical incident" question addressed to the last reading asked how the engineer *initially* found out about this article. Thus, the age of the article when last read may be several years after it has been initially identified. In fact, 36% of articles initially identified by browsing are over two years old when last read. As shown on this table, most browsing is from personal subscriptions, all of which were print in the surveys. Even though electronic subscriptions are available, no reading in the surveys of engineers revealed use of this format for browsing. Scientists also tend to continue browsing their print personal subscriptions even though most subscriptions are now available electronically.

Table 10.4   Proportion of Readings of Articles That Are Identified by Various Means by University and ORNL Engineers: U.S. 2000–2003

Means of Learning About Articles	University (%)	ORNL (%)
Browsing journal	**50.0**	**46.4**
Personal print copy	[61.8]	[46.2]
Library print copy	[11.8]	[23.1]
Library e-copy	[23.5]	[30.7]
Other digital collection	[2.9]	[*]
Searching	**17.6**	**10.7**
A&I database	[83.3]	[66.7]
OL journal collection	[8.3]	[*]
E-current contents	[8.3]	[33.3]
Another person informed	**13.2**	**28.6**
Cited in publication	**14.7**	**7.1**
Other or doesn't know	**4.4**	**7.1**
Total	99.9	99.9

*No observations.
Source: Surveys at ORNL, Drexel University, University of Pittsburgh, University of Tennessee (n = 96).

On the other hand, when library journals are browsed, the engineers will usually use electronic versions when available from the library, because it takes them less time to do so. More is said about this later.

Sometimes engineers have a problem or an information need that requires searching for information of which they were previously unaware. This is usually done by searching online (or other automated) bibliographic databases. Such searches can be done from traditional abstracting and indexing (A&I) databases or search engines such as Google. In no instance was the latter reported by the engineers surveyed. About 15% of readings by university engineers and 7% by ORNL engineers were identified in this way. Once online searching identifies a needed article, an engineer must locate where to get a copy of the article such as a library collection or some other source. Some services provide integrated online full-text journals and search capabilities, thus eliminating the additional step. While not widely available to engineers, this capability will undoubtedly become ambiguous in the future.

Often another person (such as a colleague or librarian) will inform an engineer about an article when asked or by mentioning it to the engineer. This was true for 13% of university readings and 29% of ORNL readings. Sometimes the person will also provide the engineer with a copy of the article which, as shown previously (Table 10.3), happened with 10% of the university readings and 11% of the ORNL readings. Citations are also used to learn about needed articles. In our surveys of engineers, we estimate that about 14% of university readings and 7% of ORNL readings are found through citations in other publications.

10.6 AGE OF ARTICLES READ

A majority of engineers' article readings are from recently published journals (see Table 10.5); however, an appreciable amount of reading is from articles that were published over 10 years prior to the reading.

The oldest article reported by engineers in the surveys was 33 years old. It appears that the university engineers tend to read older articles more frequently than the ORNL engineers, and it is believed that other nonuniversity engineers tend to read even newer articles than those from ORNL. Generally, older articles tend to be more valuable in that they are more important to the principal purpose for which they are read (4.84 average importance at 1 or 2 years vs. 5.22 over 2 years) and engineers tend to spend more time reading older articles: 40 minutes for articles 1

Table 10.5 Proportion of Article Readings by University and ORNL Engineers by Age of Articles Read: 2000–2003

Age of Articles (years)	University (%)	ORNL (%)
1	54.6	60.7
2	15.4	14.3
3	8.6	10.7
4,5	7.1	*
6–10	4.3	10.7
Over 10	10.0	3.6
Total	100.0	100.0

*No observations.
Source: Surveys at ORNL, Drexel University, University of Pittsburgh, University of Tennessee (n = 98).

or 2 years old, 48 minutes at 3 to 5 years, and 58 minutes for those older than 5 years.

Table 10.6 displays the sources of articles read by the age of these articles. The library collection is the principal source across all the ages of articles, but becomes the dominant source as the age increases. There are some older readings of personal subscriptions, even over five years old, although this represents only 6% of all personal subscription readings. Somewhat surprisingly, the age distribution for articles read from electronic journals is almost the same as for those read from print journals. It appears that some articles are identified by one means and then obtained from the library's electronic collection when available.

Also, as shown in Table 10.7, the electronic journals provide some older articles that are read. Somewhat surprisingly, the age distribution for articles read from electronic journals is almost the same as for those read from print journals. It appears that some articles are identified by one means and then obtained from the library's electronic collection when available.

As might be expected, newer articles are predominantly found through browsing as shown in Table 10.8. The means of identifying articles is essentially what one might expect as the articles become older.

10.7 FACTORS THAT AFFECT USE

There are many factors that contribute to the use or nonuse of journals. An extensive discussion of these factors is given in King

Table 10.6 Proportion of Article Readings by University and ORNL Engineers by Source of Articles and Their Age: 2000–2003

Age of Articles (Years)	Personal Subscriptions (%)	Library Collections (%)	Separate Copies (%)	Total (%)
1,2	37.1	40.0	22.9	100.0
3–5	21.4	42.9	35.7	100.0
6–10	33.3	66.7	*	100.0
Over 10	*	75.0	25.0	100.0

*No observations..
Source: Surveys at ORNL, Drexel University, University of Pittsburgh, University of Tennessee ($n = 98$).

Table 10.7   Proportion of Article Readings from Electronic or Print Journals by University and ORNL Engineers by Age of Articles Read: 2000–2003

Age of Articles (Years)	Electronic Journals (%)	Print Journals (%)
1,2	67.9	72.8
3,5	14.3	14.3
6–10	10.7	4.3
Over 10	7.1	8.6
Total	100.0	100.0

Source: Surveys at ORNL, Drexel University, University of Pittsburgh, University of Tennessee (n = 98).

and Tenopir (2001) and elsewhere in this book. The surveys described in this chapter have addressed some of these factors that affect journal or journal-related service use. These factors affect choices made concerning use of a particular journal, specific articles and information content, the means to identify articles, and the sources to obtain them. Some important factors related to use are:

• The awareness of particular journals, specific articles, identification means, and sources
• Ease of use, including intellectual and physical effort required to use journals and related services
• The importance of and satisfaction with attributes of journals and related services
• Availability and attributes of alternatives to the journals and related services

Table 10.8   Proportion of Article Readings of University and ORNL Engineers by How They Were Identified by Age of Article: 2000–2003

Age of Article (Years)	Browse (%)	Online Search (%)	Citation (%)	Person Told (%)	Total (%)
1,2	58.8	20.6	11.8	8.8	100.0
3–5	35.7	7.1	28.6	28.6	100.0
Over 5	*	38.5	30.8	30.8	100.1

*No observation.
Source: Surveys at ORNL, Drexel University, University of Pittsburgh, University of Tennessee (n = 97).

Some examples are given below that show how the factors influence the use of journals and so on.

Awareness does not mean that a journal or service will necessarily be used, but lack of awareness means that a needed journal service probably will not be used. While awareness of journals may seem obvious, it may be that the lack of awareness has been a problem in the past and might be in the future. In 1977, a national survey of scientists revealed that, on average, they read at least one article from 13 journals. That number has currently increased to well over 20 journals, thus indicating that there may have been a lack of awareness of needed journals in the past. The increase in the range of journals read appears to be related to a combination of emergence of electronic journals and more extensive use of online searching of bibliographic databases. Also, the general decrease in personal subscriptions and reliance on library collections may contribute to this phenomenon (but less so for engineers).

All university respondents who were asked about electronic journals indicated they were aware of them, but 15% had not used them in the past 30 days. Awareness of some new features of journals and their extent of use are summarized below:

- *Journals published exclusively in electronic format:* 89% aware, 4% have published in them.
- *Backward and forward citation links:* 70% aware, 41% have used.
- *Links to numeric databases and images:* 52% aware, 15% have used.
- *Electronic journals available back to their earliest published issues:* 85% aware, 75% have used.
- *Large preprint archives:* 63% aware, 26% have used.
- *Author websites:* 44% aware, 44% have used, 19% have their own author websites.

Thus, scientists and, perhaps to a lesser degree, engineers appear to be aware of advanced features of electronic journals and do use them, but not necessarily extensively so.

Chapter 2 provided a schema showing the many channels by which scientific and technical (engineering) information content is communicated. Journal articles are somewhat far down the stream following informal reporting, technical reports, and conference presentations and proceedings, among many other channels. Thus, engineers have substantial opportunity to be exposed to the information content found in articles that are subsequently read by them. In fact, the engineers in the surveys were asked (concerning their last article read): "Prior to your first reading of this article, did you know about the information reported or discussed in this articles?" Nearly half of the engineers reported that they were aware of the information, although they read the article anyway, sometimes at great length. One concern expressed about author websites is that the lack of awareness of specific websites is a barrier to their use; another is the continuity of them when an author moves, retires, or dies. Studies of library use have shown that sometimes engineers and other professionals are unaware of the extent of their library collections and search services, and once aware, say that they will begin using them (Griffiths and King, 1993, 1991). For example, 21% of special library users, such as those at ORNL, were unaware of online database search services and one-third of those unaware said they would likely use the service in the future (now they are aware).

Ease of use is an important factor in the extent to which journal-related services and advanced features are used. The time required to identify, locate, and obtain needed articles has clearly affected use. For example, having electronic journals available in library collections has meant they are much more extensively used than the available print collections. One reason for the use of electronic collections is that engineers (and scientists) are observed to spend about 15 minutes less time per reading to obtain the electronic version from library collections. On the other hand, with personal subscriptions, about 80% of the readings continue to be from print even though electronic versions are available. Earlier surveys had indicated it took less time to browse print journals than electronic versions, but more recent and refined data suggest that this might not be true. Nevertheless, it appears that engineers consider the personal print journals to be easier to use. Another aspect of ease of use is whether electronic articles are read on the screen or printed out to read. Of the readings from electronic articles, only 24% were read on the screen. These readings tend to be of shorter duration than those printed out and read. The observation of where the electronic articles are read (i.e., on screen or printed out) tends to be consistent with observations with respondents from other disciplines, thus suggesting that ease of use is a factor in choosing whether to read on-screen or not. It is noted that a decision to print out is also dictated sometimes by the need to keep a copy available in much the same way as photocopying from print.

Journals and related services have distinct attributes that relate to their use. For example, important journal attributes are their quality (partially determined by review and editing policies), size (i.e., number of articles, pages, etc.), number of issues, non-article features (mentioned above with awareness), available format, archive availability, price, and so on. Attributes are important as well for sources of articles, such as library collections. Examples of library collection attributes include comprehensiveness of journal collections, age of collections, location (i.e., distance to users), hours of operation, accessibility of collections (e.g., traditional, compact, or remote shelving), existence of a periodicals room, format of the collection, and collection-related services such as reference support, workstation capabilities, photocopying and workstation printing availability, and so on. Search services have attributes such as price, recall and precision of search results, and display features. All such attributes have a bearing on use that is partially determined by how important such attributes are to users and how well satisfied users are with them (Griffiths and King, 1993). However, each use may have different levels of importance and satisfaction.

Finally, there are almost always alternative sources of information provided by journals and the related services. The availability of these alternatives, awareness of them and their attributes, and their ease of use will have a significant bearing on use. For example, engineers will subscribe to a journal if it is sufficiently read (they average 26 readings per subscription) and its price warrants purchase. Otherwise, engineers will seek other sources of a journal, usually their library. In the print world, they are observed to subscribe to more journals when they are further away from their library, thus trading-off their time required to go to a library to read and the expense of purchasing journals.

11

ENGINEERING COMMUNICATION PATTERNS COMPARED WITH SCIENCE AND MEDICINE

11.1 INTRODUCTION

The unique nature of how engineers communicate is best demonstrated by comparing their communication patterns to those of others. Perhaps the most frequent distinction studied is that between engineers and members of other scientific disciplines. Many of these distinctions are covered earlier in this report, but some bear repeating here. Early classic studies by Pinelli, Allen, Taylor and others found that differences in the goals and purposes of engineers and scientists result in different patterns of communication, beginning when they are students. Many of the differences are due to fundamental differences in the disciplines—engineering is focused on production and design, while science is focused more on theory and building cumulative knowledge. Engineers also differ from scientists because they work within time constraints, are more focused on authoritative answers to specific questions, do not typically cite others' work, use libraries less, and rely less on research journals (Grigg, 1998).

An assessment of information needs of seven specific fields of science (physics, chemistry, biology, geosciences, astronomy, mathematics, and computer science) and engineering is made by Gould and Pearce (1991). This assessment is based on in-depth

Communication Patterns of Engineers. By Carol Tenopir and Donald W. King
ISBN 0-471-48492-X © 2004 Institute of Electrical and Electronics Engineers

interviews and consultations with 131 individuals who teach, conduct research, or are clearly connected with scientific research. Each field of science and engineering is discussed regarding:

- The nature of research performed
- The nature of information
- The future for information in the field
- The use of engineering information, including:
    1. Primary literature (serials, patents, technical reports, standards)
    2. Major indexing and abstracting services (print and electronic)
    3. Current awareness services and products (current research, electronic networks, conference proceedings, letters, journals, newsletters, technical reports, preprints, databases)
    4. Other electronic sources
    5. Future needs and directions (of the literature, electronic resources, interpersonal information environment, education, and data).

The results clearly show that not only are engineering and science information needs different, but all individual fields of science are unique in themselves.

Many reviews of the literature point out differences between engineers and scientists. For example, Blade (1963) discusses the nature of engineering, including aspects of creativity, research, and education. Rosenbloom and Wolek (1967) present examples of differences observed regarding engineers' and scientists' sources of information. In 1988, Allen explores some of the principal differences observed in some of his and other research projects (see the following text). He also discusses the relation between science and technology, and how knowledge diffuses over time from science to technology and finally into products. He shows how information flows among these phases of research and development, and gives some data from citation analysis to demonstrate his model of this flow. In particular, he surveys most of the significant literature regarding information-seeking processes and the factors that explain these differences.

11.2   PROFESSIONS OF ENGINEERING, SCIENCE, AND MEDICINE

Raitt (1984) summarizes differences observed between engineers' and scientists' professions. He quotes Gerstl and Hutton who summarize: "Woelfle indicates that engineers are responsible for formulating basic technical concepts, transforming them from ideas into physical entities, adapting resulting products to specific applications and evaluating their usefulness. They are thus more oriented towards applications rather than the generation of new concepts and theories leading to an increase in knowledge, which is the domain of the scientist. Whereas, science is concerned with knowledge for its own sake and with the search for truth, engineering is concerned with creating devices that will be useful. The engineer is thus the medium by which the public enjoys the fruits of scientific research" (Gerstl and Hutton, 1966).

Although Gerstl and Hutton note that engineering on a technical level is closely related to and interdependent on science, they and others point to distinct differences between science and engineering, which must be appreciated. Allen considers that engineers differ from scientists in their professional activities, their attitudes, their orientations, and even their typical family backgrounds. There is often a difference in education with the scientists going through a longer, more academic educational process, and in addition there may be a difference in their information use and communication patterns. Ladendorf (1970) believes that the striking contrasts in communication behavior patterns between scientists and engineers can be traced to fundamental differences in group organization and motivation. Scientists see themselves as belonging to an amorphous group of colleagues who share research interests and attitudes, regardless of their organizations and geographical locations. In contrast, the engineer/technologist works for organizations that are product/profit-oriented and that control the work (to create or improve products).

Allen goes further and suggests that mission-oriented organizations (as opposed to discipline-oriented) do not permit, for competitive reasons, free communication between members engaged in proprietary research and people outside the organization. Because of the nature of their work, which is not contributing to theoretical advances and an increase in general knowledge, engineers will tend more to publish their results in internal reports and memos rather than in journals read by the scientific and technical community at large. They are thus results-oriented rather than information-oriented. As a consequence of this, engineers are not closely connected to the formal communications media and thus have no real reason to read journals. Indeed "technical" journals are usually incomprehensible to them. Technologies do have their own journal systems, like science, but the literature does not cumulate and build on itself the way science does. It contains fewer references to other work and the work reported serves to document end products rather than announce theories.

King Research surveys found that the activities performed by engineers, scientists, and medical professionals are considerably different; for example, in industry and government the projected amount of time spent by engineers differs as compared to others performing various work activities. In Table 11.1 these activities

Table 11.1   Average Annual Amount (Hours) and Proportion (%) of Time Spent by Engineers, Other Scientists, and Medical Professionals in Performing Various Activities: U.S. Observed 1986–1998

Type of Activity	Engineers		Scientists		Medical	
	Hours	(%)	Hours	(%)	Hours	(%)
Primary research (data collection, observation, experimentations, etc.)	281	13.2	1,168	49.0	1,166	52.2
Engineering (design, drawings, etc.)	593	27.9	209	8.8	—	—
Technical or research support	177	8.3	173	7.3	69	3.1
Secondary or background research	142	6.7	152	6.4	191	8.6
Management, supervision, hiring	336	15.8	204	8.6	79	3.5
Finance, accounting, budgeting, etc.	37	1.7	68	2.8	26	1.2
Operations, practice	83	3.9	38	1.6	243	10.9
Marketing and sales	41	1.9	68	2.8	—	—
Professional development	135	6.3	130	5.5	125	5.6
Educating and training others	136	6.4	107	4.5	182	8.2
Other	167	7.8	67	2.8	152	6.8
Total	2,128	99.9	2,384	100.1	2,233	100.1

Source: Surveys at AT&T Bell Labs, Air Products & Chemicals, Inc., Baxter Healthcare, Eastman Chemical Co., Eastman Kodak Co., National Institutes of Health, Procter & Gamble Co. Engineers ($n = 252$), Scientists ($n = 943$), Medical ($n = 585$).

are categorized generally into primary work activities (e.g., research, engineering, technical support, management, etc.). Professionals are projected to spend an average of 2,248 hours a year working (not including vacation, sick leave, holidays, etc.). One assumes that the normal eight-hour workday results in about 1,800 hours of work a year (again, excluding vacations, sick leave, holidays, etc.). Survey respondents were asked to indicate the number of hours per year, in addition to the eight-hour work day, that they devote to work for their organization or their own professional development. This amount is estimated to be on average across all professionals, 448 hours per year or 8.6 hours per week.

Table 11.1 shows that, on the average across the three professional fields in which these data were gathered, professionals perform a wide range of primary activities. Primary research, engineering, and management consume more than 50% of the time of all three professions. These numbers break down very differently within professions, however, with engineers spending the least amount of their time doing primary research. In fact, primary research is third at 13.2%, behind engineering (27.9%, which is to be expected) and management activities at 15.8%. Both scientists and medical professionals spend much less time performing management activities. The table also shows evidence of how scattered most activities are, from the standpoint of how professionals spend at least some time on the activity and how many activities have at least 25% of their time spent on them. Engineers have the most scattered activity of all, with engineering activities just barely over the 25% mark. Other scientists and medical professionals spend a greater portion of their time focused on their primary research.

Information output of professionals is summarized in Table 11.2. Two kinds of information output are presented: (1) number of times the professionals were consulted or gave substantive advice to others, and (2) number of documents prepared or written. Engineers generally compare evenly to scientists and medical professionals for different types of oral communications, except for a marked increase in the number of presentations made both internally and externally about proposals or plans.

Engineers' written communications, however, are higher than scientists and medical professionals on almost every measure (except formal records of research). Engineers produce fewer co-authored scholarly journal articles (0.04 on average as compared to

Table 11.2   Average Annual Output by Non-University Engineers, Other Scientists, and Medical Professionals by Type of Output: Observed, 1986–1996

Type of Output	Engineers	Scientists	Medical
Oral Communication			
Substantive consultation, advice (times given)	240	227	274
Presentations about research			
Internal meetings	4.0	4.8	4.7
External meetings	0.3	1.1	2.4
Presentations about proposals/plans			
Internal meetings	15.9	8.6	10.6
External meetings	11.1	3.2	2.2
Workshops, seminars, university classes			
Internal meetings	0.4	0.9	2.0
External meetings, classes	0.3	0.7	1.1
Written Communication			
Scholarly journal articles			
Sole author	1.06	0.09	0.03
Co-author	4.31	0.95	0.09
Books			
Sole author	0.12	0.02	0.01
Co-author	0.30	0.08	—
Patent documents (over 5 years)			
Application sole author	0.25	0.11	—
Patent granted sole author	0.07	0.03	—
Application co-author	0.65	0.36	—
Patent granted co-author	0.19	0.12	—
Other publications (e.g., conference proceeding)			
Sole author	0.11	0.13	0.05
Co-author	0.60	0.32	0.09
Formal records of research, technical or administrative work (e.g., lab notes, reports)	44.2	48.2	16.3
Written proposals, plans	10.3	5.7	3.1
Other Communication			
Publications edited, reviewed, refereed	0.9	4.0	6.9
Contacts with suppliers, vendors, etc.			
Letter	1.7	5.5	16.3
Telephone	6.2	4.6	24.5
Visit	1.8	1.7	0.8

Source: Surveys at AT&T Bell Labs, Air Products & Chemicals, Inc., Baxter Healthcare, Eastman Chemical Co., Eastman Kodak, National Institutes of Health, Procter & Gamble Co. Engineers (n = 252), Scientists (n = 943), Medical (n = 585).

0.95 and 0.09) and also twice as many written proposals or plans. This seems consistent with the engineers' greater number of oral communications about proposals and plans.

11.3 COMMUNICATION CHANNELS USED BY ENGINEERS, SCIENTISTS, AND MEDICAL PROFESSIONALS

Below are some study results that compare communication channels used by engineers and scientists. Some of the data involving engineers were reported in earlier chapters. For example, Rosenbloom and Wolek (1970) surveyed more than 3,000 engineers and scientists in large corporations and from a sample of members of the Institute of Electrical and Electronics Engineers. One principal focus of the data collection was to determine communication channels used by engineers and scientists to perform their work. Respondents were asked to report their most recent instance in which an item of information proved to be useful in their work (excluding someone in their immediate circle of colleagues). Channels used are summarized in Table 11.3.

Clearly, these engineers in the 1960s relied much more on channels found in their own organization than on external ones (63% vs. 33%), and they relied more on interpersonal sources than

Table 11.3 Proportion of Instances in which Engineers and Scientists Use Various Communication Channels That Were Useful in Their Work, by Type of Channel: U.S. 1960

Type of Channels	Proportion of Instances (%)	
	Engineers	Scientists
Channels within own company		
Interpersonal		
Local source (within establishment)	25	18
Other corporate	26	9
Written media (documents)	12	6
Channels outside company		
Interpersonal (anyone outside company)	11	16
Written media		
Professional (books, articles, etc.)	15	42
Trade (trade magazines, catalogs, etc.)	11	9
	100	100

Source: Rosenbloom and Wolek, 1970.

on written materials (62% vs. 43%). Scientists' most important source was the published literature (43% of instances).

Allen (1988) reports comparisons observed in the early 1980s between communication channels used in performing technological projects and scientific research projects, summarized in Table 11.4.

These results suggest that engineers are more dependent on colleagues and scientists use the literature more than engineers do. Allen points out that engineers also need different kinds of journals and they use the literature for entirely different purposes. Engineers spend 7.9% of their time using the literature versus 18.2% by scientists (or 48% and 64% of total time spent communicating).

The amount of reading done by professionals also varies between professions, as does the type of document read (see Table 11.5). Total readings are lowest among engineers, who read only 257 documents per year on average, compared to 316 documents for scientists and 352 documents for medical professionals. The pattern of total readings is reflected in the number of scholarly journal articles read.

Internal reports are much more frequently read by engineers and they read fewer professional books. The rest of the documents appear to be read an equal amount by the engineers, scientists, and medical professionals.

Leckie, Pettigrew, and Sylvain (1996) studied and compared

Table 11.4 Proportion of Instances in Which Various Channels Were Used for Technological and Scientific Projects, by Type of Channel: U.S. 1983

Type of Channel	Proportion of Instances (%)	
	17 Technological Projects	2 Scientific Research Projects
Literature	8	51
Vendors	14	0
Customer	19	0
Other external sources	9	14
Lab. technical staff	6	3
Company research programs	5	3
Analysis and experimentation	31	9
Previous personal experience	8	20
	100	100

Source: Allen (1984).

Table 11.5 Average Amount of Reading* in a Year by Engineers, Other Scientists, and Medical Professionals by Type of Document: U.S. 1986–2003

Type of Document	Engineers		Scientists		Medical	
	Readings	(%)	Readings	(%)	Readings	(%)
Scholarly journal articles	98	36.0	148	44.4	241	66.8
Trade journals, bulletins	47	17.3	55	16.5	52	14.4
Professional books	14	5.1	28	8.4	21	5.8
Other books	26	9.6	27	8.1	23	6.4
Internal reports	73	26.8	44	13.2	12	2.3
External reports	8	2.9	13	3.9	11	3.3
Patent documents	6	2.2	18	5.4	1	3.0
Total	272	99.9	333	99.9	361	100.0

*Readings are defined as going beyond the title and abstract to the body of the document. There can be multiple readings of a particular document.
Source: Surveys at AT&T Bell Labs, Air Products & Chemicals, Inc., Baxter Healthcare, Eastman Chemical Co., Eastman Kodak Co., National Institutes of Health, Procter & Gamble Co., Oak Ridge National Laboratory, Drexel University, University of Pittsburgh, University of Tennessee. Engineers (n = 401), Scientists (n = 1,344), Medical (n = 740).

the information-seeking behavior of engineers, health care professionals, and lawyers and provide an extensive literature review of similar studies conducted in the past. They point out that the emphasis in engineering is on solving technical problems, with a resultant product, process, or service. Documentation is a byproduct, but they typically need more information than they create. The product of a scientist's work, in contrast, is often new knowledge presented in papers or reports and they use information to create more information (Leckie, Pettigrew, and Sylvain, 1996).

Lalitha compared engineering and medical professionals in India and found more similarities than differences. Engineers do write more, in particular more "semiformal" documents such as reports, and conduct more research projects. They belong to more professional organizations, but both groups are motivated by self-improvement and feel the need to keep up-to-date in their fields. Both medical and engineering professionals want current materials, but are selective in what they want. Engineers are slightly more dependent on formal sources such as books, monographs, journals, reports, patents, standards, and conference proceedings, but they tend to use only those sources that are readily available. Use and accessibility were found to be important to both groups of professionals. This study also found that informal communication is "more organized" and "more active" in medicine than in engineering (Lalitha, 1995).

Use of computers and electronic resources also varies with workfield. Saudi engineering faculty are more likely than science faculty to use computers for teaching and research (93% of engineers vs. 67% of scientists) and are twice as likely to have a computer on their desks (Al-Shanbari and Meadows, 1995). Although they do not specifically mention engineers, Kling and McKim (2000) and Mahé, Andrys, and Chartron (2000) provide detailed analyses of the different rate of acceptance of electronic media in various professional fields and conclude that widespread adoption is not inevitable in all fields, or of equal value. High- energy physicists lead in e-media adoption by sharing working papers and preprints (Kreitz et al., 1996). Molecular biologists rely more on archival journals, but share raw data through databases, and information researchers share some information electronically.

Adoption of electronic journals can be predicted by studying traditional communication patterns in workfields, as "each discipline has its own background activity patterns and information needs" (Mahé, Andrys, and Chartron, 2000). Scientists in restricted-flow fields that rely more on peer-reviewed journals (such as chemists, molecular biologists, and psychologists) are less willing to circulate electronic preprints than are those in open-flow fields, such as high-energy physicists and computer scientists (Kling and McKim, 2000). Older disciplines (such as physics, chemistry, or mathematics) need older information more than fast-changing ones (i.e., biology and computer science) (Mahé, Andrys, and Chartron, 2000).

11.4 RESOURCES AND TOOLS USED BY ENGINEERS, SCIENTISTS, AND MEDICAL PROFESSIONALS

11.4.1 Time Spent on Communication-Related Activities

Based on surveys by King Research, professionals are projected to spend about 58% of their time communicating. This information is obtained by asking professionals first to indicate the proportion of time they spend performing their primary work activities (the sum being 100%). We then ask them to subdivide their time into the proportion spent actually doing the analysis, experiments, accounting, supervision, etc. and the proportion spent communicating (e.g., through consulting or advising others, writing, reading, etc.). Table 11.6 shows engineers spend 17.7% of their time receiving information input through interpersonal means, as compared to 13% for scientists and 13.2% for medical professionals. Comparing these percentages to the time spent on information input through reading shows that engineers spend less time in reading activities than interpersonal communications. Both scientists and medical professionals spend almost two-thirds of their informational input time performing reading activities.

The data on the amount of time spent reading various types of documents reflects, to a degree, the pattern of reading given in

Table 11.6 Average Annual Amount (Hours) and Proportion (%) of Time Spent by Engineers, Other Scientists, and Medical Professionals in Performing Communication Activities: U.S. 1986–1998

Type of Communication Activity	Engineers		Scientists		Medical	
	Hours	(%)	Hours	(%)	Hours	(%)
Informal discussions	207	9.7	166.6	7.0	153	6.9
Information input						
Attending internal meetings	136	6.4	111	4.7	39	1.7
Attending external meetings	34	1.6	32	1.3	103	4.6
Reading articles, reports, e-mail, etc.	280	13.1	553	23.2	598	26.8
Total input	450	21.1	696	29.2	740	33.1
Information output						
Consulting/advising others	222	10.4	175	7.3	188	8.4
Internal presentations	99	4.7	69	2.9	21	0.9
External presentations	24	1.1	29	1.2	15	0.7
Writing • proposals, plans	92	4.3	92	3.9	30	1.3
• technical reports	117	5.5	124	5.2	33	1.5
• articles, books, etc.	12	0.6	26	1.1	125	5.6
• programs, software	17	0.7	9	0.4	7	0.3
Total output	583	27.3	524	22.0	419	18.8
Total communication time	1,240	58.3	1,386	58.1	1,312	58.9
Total work time	2,128	100.0	2,384	100.0	2,233	100.0

Source: Surveys at AT&T Bell Labs, Air Products & Chemicals, Inc., Baxter Healthcare, Eastman Chemical Co., Eastman Kodak Co., National Institutes of Health, Procter & Gamble Co. Engineers ($n = 252$), Scientists ($n = 943$), Medical ($n = 585$).

Table 11.5. Engineers spend most of their time in reading activities (30.7%) on other types of documents, including e-mail. This time spent on other readings explains the lower amount of total readings in Table 11.7.

Tables 11.5 and 11.7 also reflect a trend in reading activity by engineers, which is illustrated in Table 11.8. The table shows that the average number of scholarly article readings has remained constant over a 23-year time span, but the average number of hours spent reading these documents has fluctuated. The amount of time spent reading in 2000–01 is still an increase over the time spent reading in 1977, but is a drop from the amount of time spent in 1984.

11.4.2   Use of Tools

Griffiths et al. (1991) also report that the use of libraries for work-related purposes (late 1980s) is far less by engineers than by scientists: 39 times per year per person by engineers; 96 times by natural scientists; and 80 times by other scientists. A national survey reported by them in 1984 shows 54, 60, and 68 times per year, respectively. The late 1980s results are largely from compa-

Table 11.7   Average Annual Amount (Hours) and Proportion (%) of Time Spent Reading by Engineers, Other Scientists, and Medical Professionals by Type of Document Read: Observed 1986–2001

	Engineers		Scientists		Medical	
Type of Activity	Hours	(%)	Hours	(%)	Hours	(%)
Scholarly journal articles	81	28.0	114	20.4	285	49.3
Trade journals, bulletins	11	3.8	20	3.6	19	3.3
Professional books	19	6.6	58	10.4	66	11.4
Other books	24	8.3	39	7.0	50	8.7
Internal reports	58	20.7	53	9.5	22	3.8
External reports	6	2.1	16	2.9	24	4.2
Patent documents	4	1.4	19	3.4	2	0.3
Other (including e-mail)	86	29.7	239	42.8	110	19.0
Total	280	99.9	558	100.0	578	100.0

Source: Surveys at AT&T Bell Labs, Air Products & Chemicals, Inc., Baxter Healthcare, Eastman Chemical Co., Eastman Kodak Co., National Institutes of Health, Procter & Gamble Co., Oak Ridge National Laboratory, Drexel University, University of Pittsburgh, University of Tennessee. Engineers ($n = 401$), Scientists ($n = 1,344$), Medical ($n = 740$).

Table 11.8 Average Annual Number of Scholarly Article Readings by Engineers and Time Spent Reading: U.S. 1977, 1984, 2000–01

1977	1984		2000-01		
Readings	Time (Hours)	Readings	Time (Hours)	Readings	Time (Hours)
80	60	80	105	83	72

Source: King Research national surveys 1977, 1984; University of Tennessee, Oak Ridge National Laboratories 2000–2001.

nies and government agencies, whereas the 1984 national statistical survey included academic engineers and scientists as well. It is believed that academic engineers use libraries more frequently than do other engineers. In the 1984 survey, engineers were observed to use automated bibliographic searching far less than scientists: 0.8 average times per year by engineers; 5.8 times by natural scientists; and 1.7 times by other scientists. The 1984 survey showed that the proportion of engineers who use computers is about the same as scientists (85% of engineers), but engineers averaged fewer hours using the computers.

12

THE NASA/DOD AEROSPACE KNOWLEDGE DIFFUSION RESEARCH PROJECT

12.1 INTRODUCTION

It is unlikely that any researcher studying the communication patterns of engineers has matched the output that Thomas Pinelli and his collaborators have produced over the past 15 years. Through survey work undertaken by the *NASA/DOD Aerospace Knowledge Diffusion Research Project,* Pinelli has produced a clear picture of how aerospace engineers and scientists produce and use technical information and of the value they place on knowledge. His scope includes studies of aerospace engineers at all stages of their careers, from different types and sizes of organizations, and from numerous nationalities. As a NASA researcher, he has paid particular attention to the role that federally funded research and development plays in the work of aerospace engineering.

Pinelli and his colleagues explore the production, transfer, and use of knowledge in *Knowledge Diffusion in the U.S. Aerospace Industry.* Specifically, their work focuses on the role that federally

Communication Patterns of Engineers. By Carol Tenopir and Donald W. King
ISBN 0-471-48492-X © 2004 Institute of Electrical and Electronics Engineers

funded research and development plays in the large commercial aircraft sector of the U.S. aerospace industry. Millions of taxpayer dollars are spent on such R&D annually, yet little is known about how the resultant aerospace knowledge diffuses through the industry and elsewhere. The study aimed to understand the diffusion at the individual, organizational, national, and international levels.

12.2   FOCUS ON COMMERCIAL AIRCRAFT MANUFACTURING IN THE UNITED STATES

While the number of manufacturing firms has decreased through time, the level of international collaboration continues to grow. Today, international collaboration is the norm as the U.S. firms Boeing and McDonnell Douglas rely heavily on foreign companies to supply components. This creates serious implications for the diffusion of knowledge. Knowledge is seen as critical for the success of the U.S. aerospace, yet knowledge must be shared for international collaborations to work. As a result, knowledge must be carefully managed for U.S. firms to maintain their preeminent marketplace in the global economy.

U.S. public policy shapes the production of large commercial aircraft. Government intervention in the industry's development occurred once the military and economic values of aircraft were recognized. Pinelli et al. believe there are five factors that create government interest today:

1. The industry's importance to the economy and national security
2. The positive trade balance created by industry sales
3. The importance of tourism, dependent on large aircraft, to the economy
4. Facilitation in trade by moving people and goods
5. Growth of the industry in the job sector

The effect of U.S. public policies, whether aimed at the industry or directed toward another national objective (i.e., national security, economic well-being, foreign policy), has been to increase both pull incentives, which create demand for large commercial aircraft, and push incentives, which create supply. Generally, public policies have favored knowledge creation over knowledge diffusion with the government taking a hands-off approach. As the aircraft industry faces increasing global competition, this laissez-faire approach is being questioned as knowledge management becomes more necessary.

The federal government has had an ongoing role in aeronautical research and technology knowledge production. Public funding for science and technology is a recent development stemming from World War II and the Cold War; however, funding has been tied to either specific agencies or national goals (i.e., national security, industry development). Recent developments have seen a funding shift toward dual-use technologies (i.e., having both military and commercial potential) that would have economic benefits. Pinelli et al. (1997a,b) critique the system as it favors knowledge production over knowledge diffusion. For the U.S. aircraft industry to remain economically vital, they argue that aeronautical policy should optimize the diffusion of knowledge produced from federally funded research and technology, from foreign research, and from cooperative ventures.

NASA has had a major role in disseminating the results of federally funded aeronautic research and technology. Since 1917, the U.S. aircraft industry has been a beneficiary of research preformed by NACA (National Advisory Committee for Aeronautics) and later by NASA. Innovations created by NACA fostered the establishment of the industry, and NASA research is a major source of external knowledge used by the large commercial aircraft sector today. NASA employs both informal and formal communications to disseminate its research and technology to the air and space industries. Informal communication includes collegial contacts and personnel visits among academia, industry, and NASA. Formal communication includes the publication and presentation of NASA-funded work. The authors contend that the informal system is problematic as knowledge users can only learn what the collegial contact happens to know. Formal communication is difficult because it employs a one-way, source-to-user transmission and because it relies heavily on information intermediaries. The authors hope the system could be modified to place more emphasis on knowledge transfer and utilization and to have greater concern for the knowledge user.

12.3 COMMUNICATION BY U.S. AEROSPACE ENGINEERS AND SCIENTISTS

Like other researchers, Pinelli and his associates have found differences between engineers and scientists in their use of knowledge. Scientists are producers of knowledge (facts) and engineers are producers of designs, products, and processes (artifacts). There has been a tendency toward a convergence of scientists and engineers in the aerospace field, however, the social part of the engineers' work distinguishes them from scientists. Pinelli et al. (1997a,b) argue that the "science" and "engineering" resulting from federally funded aerospace research should continue to be published in journals and technical reports, respectively, with greater emphasis placed on engineering. They note that a substantial amount of federal engineering research remains undocumented. Finally, they hope that knowledge derived from federal research and development is diffused to meet the needs of a variety of scientists and engineers coming from a variety of organizations.

Much of the work Pinelli did was under the *NASA/DOD Aerospace Knowledge Diffusion Research Project,* which studied the diffusion of aerospace knowledge at various levels and the information-seeking behavior of aerospace engineers and scientists. Started in 1987 and lasting 10 years, the project was undertaken by researchers at the NASA Langley Research Center, the Indiana University Center for Survey Research, and Rensselaer Polytechnic Institute. The researchers examined five areas:

1. The U.S. aerospace industry
2. U.S. policies that influenced aerospace commerce
3. U.S. technologies policies and their relation to the aerospace industry
4. The system used to transfer federally funded aerospace research and development
5. The relationship between science and technology

The project, which consisted of self-reported mail questionnaires and telephone surveys, was conducted in four phases. Phase 1 focused on the information-seeking behavior of U.S. aerospace engineers and scientists. Survey respondents were drawn from the American Institute of Aeronautics and Astronautics (AIAA), the

American Society for Testing and Materials (ASTM), the Human Factors and Ergonomics Society (HF&ES), the Society of Automotive Engineers (SAE), the Society of Flight Test Engineers (SFTE), the Society for Advancement of Material & Process Engineers (SAMPE), the Society of Manufacturing Engineers (SME), NASA aerospace technologists (ASTs), and subscribers to *Aerospace Engineering* and *Manufacturing Engineering.*

Phase 2 explored the role intermediaries play in the diffusion of aerospace knowledge. Survey respondents were drawn from several sources, including the *Directory of Special Libraries and Information Centers* and the Special Libraries Association. Phase 3 studied the information-seeking behavior of U.S. aerospace students and faculty and the role played by academically affiliated intermediaries. Survey respondents were drawn from faculty and students involved with either the NASA/USRA Capstone Design Program, aerospace programs accredited by the Accreditation Board for Engineering and Technology (ABET), or the AIAA. Phase 4 looked at the information environment of non-U.S. aerospace engineers and scientists. Site-specific surveys were performed in India, Israel, Japan, the Netherlands, and Russia.

In Phase 1 surveys, a majority of all respondents labeled themselves as engineers (except in the AIAA Survey 10). Day-to-day responsibilities varied among the survey population, as did education levels. Scientists and engineers from academia, government, and industry were all represented. The survey pool was divided into nine aerospace-related disciplines. The majority of respondents in all disciplines were male, with propulsion and production having the highest percentage of men (97%) and avionics the lowest (76%).

Respondents from all disciplines believed that effective communication is important to the success of their duties. Hours spent communicating per week was high in all disciplines, with a low of 27.8 hours per week for research to a high of 46.0 for sales and service. More time is spent communicating information to others than working with information received from others. With the exception for avionics and design, respondents in all other disciplines spend more time communicating orally than they do in writing. Other trends that emerged were that time spent communicating has increased over the last five years and, as respondents advance professionally, they spend more time communicating.

Most of the respondents had been involved in collaborative writing over the last six months, with a high of 100% for production and a low of 67% for flight test. Across all disciplines, the respondents produced more informal information products (i.e., letters) than formal products (i.e., journal articles). Letters and memoranda were produced most of all. They also used more informal than formal products. Respondents tended to be higher users of information products than producers of products.

Pinelli et al. examined the use and importance of five technical information channels: (1) conference and meeting papers, (2) journal articles, (3) in-house technical reports, (4) DOD technical reports, and (5) NASA technical reports. Respondents in research and avionics tend to use and regard as important these products more than do designers and scientists in the other disciplines.

Eight factors affecting the use of selected information channels were also examined. These eight factors were categorized into three groups: (1) accessibility, (2) cost, and (3) content. For conference papers, journal articles, in-house reports, and NASA reports, five factors were the most important: relevance, technical quality, comprehensiveness of data and information, ease of obtaining, and ease of reading and using. Of all the factors, relevance was chosen as most important when deciding whether to use an information channel. Across the disciplines there was little variation in the ranking of factors.

Using the same eight factors, the respondents gave their opinions of the information channels. For conference papers, the respondents assigned "relevance to work" and "good prior experience with use" as their highest ranking. For journal articles, "good technical quality" and "good prior experience" received the highest marks. For in-house technical reports, the highest ranks were given to "easy to obtain physically," "good prior experience using them," and "can be obtained at nearby location or source." For NASA technical reports, respondents rated "good technical quality," "comprehensive data and information," and "good prior experience using them" as highest.

The majority of respondents used computer networks in their professional duties with a high of 92% for design and a low of 76% for sales and service. Across all disciplines, respondents considered computer networks important for performing their jobs. Networks were used most often for electronic mail and connecting to geographically distant sites. They were used least often to order documents from the library and to prepare papers with colleagues at geographically distant sites.

Respondents were asked questions concerning the use and importance of libraries. Most respondents indicated that their organization had a library. Respondents in human factors, avionics, production, and sales and service had the highest library use rates in the survey. Respondents in research considered the library most important for performing their professional duties. Of those respondents who had not used a library in the past six months, the most common reason was that their information needs were easily met some other way.

Survey respondents were asked to identify the information-seeking steps they take to find information needed to solve a professional task. The typical search strategy was to:

1. Search a personal store of technical information.
2. Speak with co-workers inside the organization.
3. Speak with colleagues outside the organization.
4. Use literature resources in the organization's library.
5. Search an electronic database.
6. Consult a librarian.

Pinelli et al. (1997a,b) found that U.S. aerospace engineers and scientists who used federally funded research and design tended to get it from multiple sources. Across all disciplines, respondents used co-workers and colleagues inside and outside their organization and NASA and DOD technical reports at about the same rate to find out about federally funded R&D. Respondents in sales and service, research, and avionics had the highest use of NASA and DOD contacts. NASA and DOD workshops, visits to NASA and DOD facilities, and publications such as *NASA STAR* were the least-used methods to find out about federally funded R&D.

In a separate survey of AIAA members, the authors examined the use, frequency of use, and familiarity with bibliographic products that provide access to federally funded R&D. Respondents were questioned about four print and three electronic bibliographic products. Overall response rates indicate that little use is made of these products. *NASA STAR* had the highest use rate, but still less than 25% of the respondents consulted it. Fewer than 10% of the respondents used the other print sources. The majority of the respondents were not familiar with the products, in particular with the electronic sources. Among respondents who were familiar with the print products the major factors in non-use were lack of availability and accessibility and reliance on others to search for information. For the electronic sources, the major reason for non-use was they could get the information more easily from other sources.

Pinelli et al. (1997a,b) also surveyed members of the AIAA, the SAE, and the SME to determine how uncertainty affects information-seeking behavior and the use of federally funded aerospace research. The AIAA respondents on average had more schooling and professional experience. The SME survey had the smallest number of respondents who were educated as engineers. A majority of AIAA and SAE respondents were involved in design and development, while the majority of SME respondents were involved in manufacturing and production.

Respondents from the three surveys all reported similar patterns of information seeking. Informal and internal information channels and sources were favored first. If the information needs could not be met, formal and external information sources and channels were then used. Technical uncertainty influenced the selection of information channels and sources; however, it was not a strong predictor of the frequency of use. There was also no strong correlation between technical uncertainty and the amount of time spent communicating.

As technical uncertainty increased, there was a higher use by respondents of federally funded aerospace R&D found in DOD and NASA technical reports. Factors that contributed to the use of these technical reports included technical quality, comprehensiveness, and relevance. Respondents rated cost and ease of access as less important.

The communication practices and information activities of new U.S. aerospace engineers and scientists are revealed in a survey of AIAA members who had recently changed their status from student to professional. Pinelli et al. compared the responses of the new engineers to those of engineering students and established engineers.

Respondents were asked to rank engineering, science, and management goals and aspirations. Both new engineers and engineering students gave the highest ratings to engineering goals. However, engineering students ranked science and management goals as more important than the new engineers.

Established engineers rated the importance of communicating technical information effectively much higher than the new engineers. Established engineers also spent more time communicating and working with technical information than the new engineers. Both groups reported that they wrote in collaboration, and both new and established engineers considered collaborative writing as more productive than writing alone.

New engineers and engineering students were asked to rank the importance of various communication and information-use skills to future personal success. Both groups rated these skills as important, with higher ranks coming from the engineering students. At least 50% of the respondents had received instruction in communication skills, with engineering students generally rating the helpfulness of the instruction higher than the new engineers.

Library use was nearly equal for both new and established engineers. The most common reasons for non-use by both groups were "My information needs were more easily met some other way" and "I had no information needs." Thirty-five percent of both groups had not used a library to complete their most important project and a majority of respondents from both groups had not consulted a librarian on their most important project.

For completing the most important tasks of key problems, new engineers relied more on colleagues and co-workers than established engineers as part of their information search. Established engineers relied more heavily on their personal stores of knowledge. Both groups relied on colleagues and co-workers to find out about federally funded aerospace R&D, while NASA and DOD were the most important written sources for both groups to find out about that type of R&D. Both groups considered the time it took to locate and obtain the results of federally funded R&D to be a problem.

The authors also studied the career choice, satisfaction, expectations, and the technical communication practices and instruction of U.S. undergraduate and graduate engineering students. The work was based on a survey of AIAA student members and represents a Phase 3 activity of the *NASA/DOD Aerospace Knowledge Diffusion Research Project.*

Undergraduates placed high importance on achieving career success through advancement within an organization. Graduate students placed high importance on developing a professional reputation through communication. These goals reflect the former's leanings toward industry and the latter's preference toward academia.

Both undergraduate and graduate students believed that communication and information-use skills are important for professional success. The majority of both groups have received instruction in these skills. Undergraduates appeared to be as competent as graduate students in information-seeking skills. However, undergraduates used these skills less often. Fifteen percent of the undergraduates claimed not to have used a library during the school term, compared to only 5% of the graduate students.

Undergraduate and graduate students differed in the sequence of sources they consulted when problem solving, but both groups relied nearly half the time on their own stores of personal or technical information as the first step. Nearly three-quarters of the undergraduates (74%) and two-thirds of the graduate students (67%) did not consult with a librarian at all during the search process.

Sources of information used varied between the undergraduates and graduate students. Undergraduates had a preference for textbooks, handbooks, audiovisual materials, and drawings or specifications. Graduate students had a preference for journal articles, computer programs and documentation, conference or meeting papers, theses or dissertations, U.S. government and industry technical reports, and technical translations.

A Phase 4 activity of the *NASA/DOD Aerospace Knowledge Diffusion Research Project* surveyed aerospace engineering professionals and students in India, Japan, the Netherlands, Russia, and the United Kingdom. Aerospace engineers and scientists showed many similarities irrespective of nationality. All respondents reported spending a large and similar amount of time communicating and working with received technical information. Most tend to write both alone and in collaboration. Collaborative writing is viewed by respondents as being equally or more productive than writing alone. The respondents as a group use more technical information than they produce, with journal articles being the most popular. With the exception of the Russians, all respondents rely heavily on personal stores of knowledge when problem solving. Speaking with co-workers is also an important source of information for all groups. Library use varied, with Indians making the most visits to libraries and Americans the fewest. Americans had the greatest access to computer networks while

Indians had the least; however, the mean importance rating of electronic networks was similar for all groups.

All students shared similar career goals and aspirations, with "having the opportunity to explore new ideas about technology or systems" rated the highest. All respondents agreed that communication and information-use skills are important for professional success. U.S. students received the most technical communication training; Japanese students received the least. The majority of respondents found collaborative writing equally or more productive than writing alone. For information needed in problem solving, all groups relied heavily on personal stores of technical information as well as talking with fellow students and faculty. Indian students had the highest rates of library use with American students having the lowest. British students, closely followed by American students, had the highest use of electronic networks, while Russian students had the lowest use.

U.S. aerospace engineering faculty were also studied by Pinelli et al. as a Phase 3 activity of the *NASA/DOD Aerospace Knowledge Diffusion Research Project*. This work was based on a survey of AIAA members who identified themselves as educators. To determine the effects of academic work experience, the sample pool was divided into those with less than 15 years of academic experience and those with more than 15 years. Both groups agreed that communicating technical information effectively is important. Both spent roughly the same amount of time communicating information as well as working with information received from others. The majority of respondents indicated that journal articles and conference and meeting papers were the most-used information products. The more experienced faculty tended to use letters and memoranda to a greater extent. Conference and meeting papers and journal articles were also heavily produced by both groups. Almost all respondents used computer networks and made use of electronic mail and the World Wide Web. Both groups of faculty find libraries to be important in performing their duties. Proximity of the library to the faculty does appear to influence use, however.

Out of five technical information products, journal articles and conference and meeting proceedings were used the most by faculty in both groups and were considered the most important. NASA technical reports were the third most frequently used and considered third most important. The most important factor regarding the use of these three products was their relevance to work.

For problem solving, faculty in both groups preferred to rely on their own personal stores of knowledge as the first step. This was generally followed by consulting with co-workers and colleagues. Over half the respondents in both groups do not consult with a librarian when problem solving.

The survey of AIAA members also produced information about the communication environment of small, medium, and large U.S. aerospace organizations. Face-to-face communication was employed most often by respondents in all three sizes of enterprises and it was also considered the most important type of communication. The majority of respondents considered written communication to be the most accurate, as well as the most useful, form of communication. In all three sizes of companies, face-to-face communications was the most frequently used way to obtain information within a department. Face-to-face conversations were most frequently used within the department to provide information; telephone conversations were most frequently used to provide information outside the department.

Overall, 80% of the respondents used electronic networks in their professional work, but half of the respondents considered the computer networks to be neither important nor unimportant. Respondents used electronic networks more frequently for internal rather than external communication.

Pinelli et al. also investigated the use of computer networks in a survey of subscribers to *Aerospace Engineering*. Respondents included engineers and scientists employed in academia, government, and industry, a majority of which had access to either a local or organizational computer network. Half had access to an external research network.

Respondents used computer networks most frequently for electronic mail, file transfer, remote log-in to access files, remote log-in to run a program, and electronic bulletin boards. Respondents used networks more frequently to communicate with co-workers than with people outside the organization. Computer networks were used for a wide variety of workplace tasks, the most frequent of which was performing mathematical analyses.

The role played by intermediaries in the diffusion of aerospace knowledge was studied in a survey of academic and aerospace industry libraries as part of Phase 2 and 3 activities of the *NASA/DOD Aerospace Knowledge Diffusion Research Project*.

Libraries and librarians play a vital role in providing NASA and DOD technical reports to their intended users. Pinelli et al. (1997a,b) found that intermediaries have a grasp of the technical information needs of their user community, but they do not play an active role in disseminating NASA-generated knowledge. Part of the problem lies in funding and staffing issues for libraries; however, the larger problem appeared to be NASA's inability to involve intermediaries in the knowledge-transfer process. The survey showed that there were few contacts between NASA and libraries regarding the transfer of federally funded R&D.

Libraries have a major role in providing access to U.S. government technical reports and Pinelli et al. also examined how these reports are used. Nearly all the respondents (97%) to a survey of AIAA members used U.S. government technical reports in their work. All groups found government technical reports to be important information products. Respondents used an average of 11.5 government technical reports over a six-month period. The main factors affecting the use of U.S. government technical reports were relevance and technical quality or reliability. The major use of the reports by the respondents was for research followed by education and management.

Nearly two-thirds of the respondents (64%) used the U.S. government technical reports in completing their most important project or solving their problem. Personal stores of knowledge were the major way respondents used to find out about government technical reports. This was followed by asking co-workers inside the organization and then by asking colleagues outside the organization. Over two-thirds of the respondents (67%) used U.S. government technical reports through the entire project or problem, while 42% of the respondents used them at the beginning. Overall, AIAA members found U.S. government technical reports both effective and efficient in helping them complete their projects and solving their problems.

12.4 OTHER RELATED WORK

Barclay, Pinelli, and Kennedy (1993, 1994) compared the technical communication practice of Dutch and U.S. aerospace engineers and scientists as part of the *NASA/DOD Aerospace Knowledge Diffusion Research Project*. The amount of time spent producing and communicating technical information was comparable for both the Dutch and Americans. High percentages of both groups (90%) considered the ability to communicate effectively as important for professional success. A majority of respondents found that the time they spent communicating had increased over the past five years.

Both groups tended to produce the same types of information products, regardless of whether work was done alone or in collaboration. Abstracts, journal articles, conference and meeting papers, letters, and drawings/specifications were the products most used by the Dutch respondents. Memoranda, journal articles, letters, conference and meeting papers, and abstracts were the products most used by the American respondents. A much greater percentage of the Dutch scientists and engineers indicated that a library or technical center was located in the building where they worked than did the Americans (44% of the Dutch compared to only 9% of the Americans). Not surprisingly then, the Dutch visited their library or technical center more than their American counterparts in the last six months.

Both groups displayed similar patterns when selecting an information source for problem solving—Dutch and American aerospace engineers and scientists relied heavily on personal stores of technical information and contacting co-workers within their organization. Nearly all the American and Dutch respondents used computers to produce technical information (98% and 91% respectively). Word processing programs were the most popular computer applications used by both groups, while 89% of Americans used electronic networks at work compared to only 65% of the Dutch. Familiarity may engender trust, as U.S. respondents considered electronic networks almost twice as important as the Dutch did.

Both groups relied heavily on their own national technical reports. The Dutch tend to have better access to foreign technical reports than the Americans (perhaps a reflection of the proximity of the libraries or maybe just a function of the relative size of the countries). Foreign-language fluency was much greater with the Dutch respondents than with the American respondents.

Pinelli, Kennedy, and Barclay (1994) investigated the importance of technical communications and information-use skills to engineering students, the instruction they received in communication skills, and the helpfulness of that instruction. This research was based on a survey of AIAA student members that was undertaken as part of the *NASA/DOD Aerospace Knowledge Diffusion Research Project.*

Students ranked the importance of six communication and information-use skills to their future professional success. Nearly all of the respondents (90%) ranked using computer, communications, and information technology as being important. These were followed by technical writing/communication (84%), speech/oral communication (84%), using engineering/science information resources and materials (80%), using a library that contains engineering/science information resources and materials (64%), and searching electronic databases (51%). At least half of the respondents had received training in one of the six skills, the high being 83% for using computer, communications, and information technology and the low being 50% for searching electronic databases. The respondents often did not consider that the instruction was helpful, however. Instruction in using computer, communications, and information technology was thought to be most helpful (68%), while instruction in using a library than contains engineering/science information resources and materials was thought to be least helpful (39%).

Pinelli et al. (1995) also examined the technical communication abilities, skills, and competencies of U.S. aerospace engineering students along with survey data on career goals and aspirations and computer use. Respondents had goals that were more engineering oriented rather than science and management oriented. Eighty-four percent of the respondents hoped to have the opportunity to explore new ideas about technology or systems and 70% desired to work on projects that require learning new technical knowledge. Over two-thirds (68%) of the students owned a personal computer. Nearly all (almost 99%) indicated that they use computers to prepare written technical communications and 82% of them always use them when writing these communications. The top two reasons given for not using a computer were "no or limited computer access" and "lack of knowledge and skills in using a computer."

Pinelli et al. (1997d) compared the technical communication practices of Japanese and U.S. aerospace engineers and scientists as a Phase 4 activity of the *NASA/DOD Aerospace Knowledge Diffusion Research Project* based on a survey conducted in Japan and the United States. Some differences were found. Japanese had a greater language fluency than

Americans—Japanese were fluent in English and Japanese and a majority read German, while Americans were fluent only in their native English. Japanese respondents spent more time writing communications, while the Americans spent greater time communicating orally. The American respondents devoted more hours to working with written and oral communication than did the Japanese. A majority of both Japanese and American respondents wrote in collaboration and believe that collaborative writing is about as productive or more productive than writing alone. Most respondents thought undergraduate aerospace students should take a class in technical communication, although this opinion was more strongly held by the Americans. Japanese respondents used their organization's library much more than their American counterparts, although both groups thought the library was important for performing their job. A much greater number of the Americans (89%) used electronic networks than did the Japanese (55%). The Americans considered electronic networks to be more important in performing their duties than did the Japanese. Both groups made high use of the NASA technical reports.

These technical reports produced by authors at the NASA Langley Research Center were studied further by Pinelli, Barclay, and Kennedy (1994), who examined reader preferences. The research was a Phase 1 activity of the *NASA/DOD Aerospace Knowledge Diffusion Research Project* and consisted of surveying producers and users of the technical reports.

Respondents indicated that the most common reading sequence of the reports was the conclusion, results and discussion, title page, introduction, and summary. Respondents used the abstract, conclusions, title page, and introduction to determine whether to read the full technical report. Both producers and users indicated that the foreword and preface could be deleted from reports. A strong majority of producers and a majority of users thought that only longer reports needed a table of contents and a majority of respondents wanted technical reports to contain a summary as well as an abstract. A majority also preferred references to be cited by number rather than by author and year. Producers tended to prefer the passive voice, while users favored the active voice. Both groups preferred reports to be written in third person voice to those written in first person.

12.5 SUMMARY

Much of our understanding of the communication patterns of scientists and engineers in the recent decades can be traced to the ongoing and systematic research of Thomas Pinelli. His work focuses on multiple aspects of communication, including information seeking, information use, and information outputs, leading also to conclusions on how information can be better structured or provided. He has frequently collaborated with others over the years, spawning a new generation of researchers.

The work of Pinelli and his collaborators paints one picture of the communication patterns of engineers that has remained stable. Engineers have unique communication patterns. They favor informal and oral communication and also use a wide variety of written sources and channels of communication, but they prefer easily accessible information sources. They value high-quality information but are not always successful in getting it or using it in an efficient manner. Engineers also must create information outputs but are not always comfortable doing so. The ongoing research of Pinelli and his collaborators also reveals changes in the 1990s that are putting new communication pressures on engineers. International cooperation, the increased need for collaboration, and advances in information technologies make information access, use, and creation an increasingly important aspect of engineers' work.

13

SUMMARY

It has been said that "communication is the essence of science" and, as a field of science, this statement certainly holds true for engineering. However, it is abundantly clear that engineers communicate differently from other fields of science. Engineers tend to rely much more on interpersonal and informal means of communication than other scientists who read journals more frequently and are most inclined to use other formal means of communication as well. The reasons for this may be the nature of engineering work, but also engineers' inherent personalities, ways of addressing problems, and learning style may play a role as well. Engineers tend to be self-sufficient and more direct in their approach to work. Their learning style emphasizes listening and discussing rather than observing and reading. It may be that those who enter engineering as a profession may lean toward a certain personality, way of thinking, and learning style.

Communication requires considerable resources, particularly time. Studies over the years indicate that engineers spend more than half their time communicating and that amount of time appears to be increasing. Yet, the quality of communication may be deteriorating, which is disconcerting because of the positive impact of good communication.

Over the years, studies have revealed many indicators of the

Communication Patterns of Engineers. By Carol Tenopir and Donald W. King
ISBN 0-471-48492-X © 2004 Institute of Electrical and Electronics Engineers

usefulness and value of communication and information use, including:

- Engineer productivity and indices of the amount of communication are correlated.
- Engineers with work that has been formally recognized (e.g., by an award) tend to communicate more than those whose work hasn't.
- Communication skills of engineers lead to better hiring opportunities and career advancements.
- Good communication results in higher quality work, faster performance, and ends up saving money and other resources.
- Communication is important to all phases of projects.

Communication is essential to life-long-learning. Science information doubles about every 15 to 20 years; that is, since 1985 all scientific knowledge discovered throughout history has now doubled. This means that an engineer at graduation will have had access to only one-sixth of the knowledge that must be mastered during a career. Furthermore, engineers in industry tend to be assigned new projects that require unfamiliar knowledge. Thus, engineers must continue to learn and substantial communication is necessary for the learning process.

Engineering communication is extremely complex, due in part to the many types of activities performed by engineers: research, design, development, production, construction, teaching, management, marketing and sales, and so on. Each of these activities relies on communication and information as a resource to perform the activity and the output of the activity is often information that needs to be recorded and communicated to others.

The many means by which engineers communicate are referred to as *channels*. Channels are both written or recorded (as databases or images) and interpersonal or oral. Examples of written or recorded channels include:

- Formal publications such as scholarly and trade journals, books, internal and external reports, patent documents, conference proceedings, standards, regulations, dissertations, and so on.
- Letters, memorandum, proposals, specifications, and so on.

- Numeric and bibliographic databases, graphics, drawings, computer programs, and so on.

Examples of interpersonal or oral channels include informal discussions, program reports, internal presentations, conference presentations, and sales presentations.

Within each channel are *sources* of information that engineers can use. For example, sources of journals include personal subscriptions, library collections and separate copies of articles such as preprints, reprints, interlibrary loan, document delivery, copies from colleagues or authors, and author websites. Each source might be available in an alternative *format*. Again with journals, library collections are printed issues (or bound), electronic, and microfilm. The complexity of communication becomes even greater because specific information can be obtained from several channels such as personal contact, conference proceedings, journal articles, and so on; although the information often appears in specific channels at different periods of time. For example, information can appear sequentially in a conference presentation, proceeding and articles over several years. There has been a recent evolution of online engineering books and reference tools that are essential for engineering undergraduate students to learn engineering concepts.

Because new knowledge continues to grow exponentially, engineers feel the need to keep up with this growth and often do so by browsing trade and scholarly journals, attending conferences and other meetings, and holding discussions with colleagues. However, there are times when work requires learning about new information, determining which channels can be used, and obtaining it from accessible sources. We refer to these communication processes as *information seeking*. Engineers' information seeking involves all the channels mentioned above, varies with each information need, and generally differs by engineering discipline (e.g., aeronautical compared with civil engineering), the nature of work being performed (e.g., research vs. design), country (e.g., based on access to technology, funds available, and culture), and personal characteristics (e.g., gender and age). What tends to be common to all engineers is the need to obtain information quickly with as little effort as possible. This need dictates to a large degree, the information seeking behavior of engineers.

For several decades there has been widespread concern over the quality of engineers' communication. Much of this concern involves writing ability that goes beyond improving grammar, sentence structure, or avoiding jargon, to clearly presenting ideas, writing to a specific audience, and writing for specific types of publications. Another concern is with interpersonal communication and with formal presentations including use of relevant graphics and articulation of central ideas. Finally, there is a sense by some communication researchers that engineers don't fully utilize communication resources such as technologies and libraries. There appears to be a link in engineers' reluctance to change and adaptation to new technologies. Libraries sometimes aren't used by engineers because they are unaware of important services that are provided or because they don't fully appreciate the benefits of their use.

Suggested solutions for improving engineers' communication include: improving education and training, organizing to facilitate communication, modifying information services, and designing flexible systems to address the complexities of engineers' communication. These approaches are as follows:

- The engineering and communication educational communities have together taken a serious approach to improving all facets of engineering communication: writing of all types of documents, oral presentations, arguing and articulating different points of view, team communication, and better utilization of various channels, sources (particularly libraries), and available technologies. Various studies and experiments throughout the world have shown substantial progress and promise.

- Engineers' organizations have taken steps to internally organize to promote better communication through recognizing and aiding information-intensive staff members, such as gatekeepers, who support other engineers; improving awareness of information channels, sources and services; train engineers to communicate better; improving internal information facilities infrastructure, and services (i.e., libraries, databases, networks, etc.); and structuring the organization and management to facilitate better communication.

- External information services such as engineering societies, publishers, information analysis centers, clearinghouses, library consortia and aggregators, bibliographic database services, and Web-based services have all been much more sensitive to engineers' communications needs and have adjusted to accommodate such needs, albeit somewhat slowly in some cases.

* A related solution to communication problems is to design flexible communications systems that will deal with the many complexities of engineering communication.

The library community is also supporting these efforts. The Library Division of the American Society for Engineering Education has recently taken an initiative to develop literacy standards for engineers.

Engineers use a variety of informal and formal, oral and written information resources to help them accomplish their tasks more efficiently and more effectively. Although information is essential to the engineering workplace environment, it is not yet optimally integrated into work projects and workday tasks. Creating computer-based information systems that capture information at all stages of the design and engineering processes can improve information use in the workplace. Electronic journals that are designed with the needs of the engineers in mind can be an important component of this system. Use of such systems, however, will be predicated on designing systems that match the way engineers work and incorporating a program of information literacy education in the workplace.

The community of engineering practitioners, faculty, students, and information specialists saw many steps taken in the 1990s towards using technology for improving information seeking by engineers and some improvements have been achieved as a result. However, the possibilities technology offers to the engineering community has yet to be fully exploited and much remains to be accomplished. Using technology more wisely can be facilitated by forums at national meetings where engineering practitioners, educators, librarians and others can discuss such issues and resolve them in a meaningful way.

In this book we have tried to show how engineers communicated in the past, why they communicated in this way, and what can be done to improve communications. We also presented examples from our studies, as well as those of others to illustrate important aspects of communication. It is hoped that these studies provide an understanding of communication patterns of engineers and that such understanding will help improve communication and its consequences in the future.

BIBLIOGRAPHY

Abels, Eileen G., Peter Liebscher, and Daniel W. Denman. 1996. Factors that influence the use of electronic networks by science and engineering faculty at small institutions, Part I: Queries. *Journal of the American Society for Information Science* 47(2): 146–158.

Accreditation Board for Engineering and Technology. 2000. Available from World Wide Web: <http://www.abet.org>.

Ackoff, Russell L. 1971. Systems, organizations and interdisciplinary research. In *Systems Thinking: Selected Readings.* F. E. Emery, ed., pp. 330–347. New York: Penguin.

Ackoff, R. L., T. A. Cowan, W. M. Sachs, M. L. Meditz, P. Davis, J. C. Emery, and M. C. J. Elton. 1976. *The SCATT Report: Designing a National Scientific and Technological Communication System.* Philadelphia: University of Pennsylvania Press.

Adams, E. B., and S. A. Rood. 1978. Critical issues in scientific and technical communication perspectives of users, providers and policymakers. Final report 1971–1978. NTIS. PB 279 382 GWU/SCD-78/1.

Adams, Judith A., and Sharon C. Bonk. 1995. Electronic information technologies and resources: Use by university faculty and faculty preferences for related library services. *College and Research Libraries* 56(2): 119–131.

Advisory Group for Aerospace Research and Development. 1998. *An International Aerospace Information Network.* AGARD/AR–366. Neuilly-sur-Siene: AGARD.

Al-Rawas, A., and S. Easterbrook. 1996. Communication problems in requirements engineering: A field study. Presented at *Professional Awareness in Software Engineering.*

Al-Shanbari, H., and A. J. Meadows. 1995. Problems of communication and information-handling among scientists and engineers in Saudi universities. *Journal of Information Science* 21(6): 473–478.

Allen, Bryce L. 1991. Cognitive research in Information science: Implications for design. In *Annual Review of Information Science and Technology* 26. Martha Williams, ed., pp. 3–37. Medford, NJ: Learned Information, Inc.

Allen, Robert S. 1991. Physics information and scientific communication: Information sources and communication patterns. *Science and Technology Libraries* 11(3): 27–38.

Allen, Thomas J. 1964. *The Utilization of Information Sources During R&D Proposal Preparation.* Report no. 97-64. Cambridge: Sloan School of Management, Massachusetts Institute of Technology.

Allen, Thomas J. 1966a. The differential performance of information channels in the transfer of technology. *MIT Sloan School of Management WP* 196–66.

Allen, Thomas J. 1966b. Managing the flow of scientific and technical information (Ph.D.). Massachusetts Institute of Technology, Available from NTIS: PB174440.

Allen, Thomas J. 1966c. Studies of the problem solving process in engineering design. *IEEE Transactions on Engineering Management* 13(2): 72–83.

Allen, Thomas J. 1968. Organizational aspects of information flow in technology. *ASLIB Proceedings* 20: 20.

Allen, Thomas J. 1969. Information needs and uses. In *Annual Review of Information Science and Technology* 4: 3–29. Chicago: Encyclopaedia Britannica.

Allen, Thomas J. 1970. Roles in technical communication networks. In *Communication Among Scientists and Engineers,* pp. 191–208. Lexington, MA: Heath Lexington Books.

Allen, Thomas J. 1976. The importance of direct personal communication in the transfer of technology. In *The Problem of Optimization of User Benefit in Scientific and Technological Information Transfer,* 2-1–2–10. AGARD.

Allen, Thomas J. 1984. *Managing the Flow of Technology: Technology Transfer and the Dissemination of Technological Information within the R & D Organization.* Cambridge, MA: MIT Press.

Allen, Thomas J. 1988. Distinguishing engineers from scientists. In *Managing Professionals in Innovative Organizations: A Collection of Readings,* pp. 3–18. Cambridge, MA: Ballinger Publishing Company.

Allen, Thomas J. 1994. *Information Technology and Corporations of the 1990s.* Oxford University Press.

Allen, Thomas J., and S. I. Cohen. 1966. Information flow in an R&D laboratory. *MIT Sloan School of Management WP* 217–66.

Allen, Thomas J., and S. I. Cohen. 1969. Information flow in research and development laboratories. *Adminstrative Science Quarterly* 14: 12–19.

Allen, Thomas J., and Peter G. Gerstberger. 1964. *Criteria for Selection of an Information Source.* Cambridge, MA: MIT Sloan School of Management.

Allen, Thomas J., Diane B. Hyman, and David L. Pinckney. 1983. Transferring technology to the small manufacturing firm. A study of technology transfer in three countries. *Research Today* 12: 199–211.

Allen, Thomas J., and Kumar S. Nochur. 1992. Do nominated boundary spanners become effective technological gatekeepers? *IEEE Transactions on Engineering Management* 39(3): 265–269.

Allen, Thomas J., and Alfred P. Sloan. 1970. Communication networks in R&D laboratories. *R&D Management* 1: 14–21.

Allen, Thomas J., et al. 1968. *The Problem of Internal Consulting in R&D Organizations.* Working paper, pp. 319–68. Cambridge: Massachusetts Institute of Technology.

Allen, Thomas J., et al. 1971. The internation technological gatekeeper. *Technological Review* 73(5): 36–43.

Allen, Thomas J., et al. 1980. R&D performance as a function of international communication project management, and the nature of the work. *IEEE Transcripts on Engieering Mangagement* EM-27: 2–12.

Aloni, Michaela. 1985. Patterns of information transfer among engineers and applied scientists in complex organizations. *Scientometrics* 8(5–6): 279–300.

Amare, N. 2002. The culture(s) of the technical communicator. *IEEE Transactions on Professional Communication* 45(2): 128–132.

American Management Association. 1970. Statistical concepts for managerial decision-making.

American Psychological Association. 1968. *Reports of the American Psychological Association's Project of Scientific Information Exchange in Psychology* 1 (1963), 2 (1965), Supplement (1968).

Ames, James. 1996. The industrial engineer and the World Wide Web. In *Proceedings of the 1996 International Industrial Engineering Conference,* pp. 158–163. Norcross, GA: IIE.

Anderson, Claire J., Myron Glassman, R. Bruce McAfee, and Thomas Pinelli. 2001. An investigation of factors affecting how engineers and scientists seek information. *Journal of Engineering and Technology Management* 18(2): 131–155.

Anthony, L. J., et al. 1969. The growth of the literature of physics. *Reports on Progress in Physics* 32(6): 709–767.

Anzieu, D. 1965. Les communications intra-groupe. In *Conference on Communication Processes.* F. A. Geldard, ed., pp. 109–188. London: Macmillan.

Ardoino, J. 1964. Information et communication dans l'enterprise et les groupes de travail. *Les Editions d'Organization 2nd Edition.*

Arechavala-Vargas, Ricardo. 1985. The communication network structures of research and development units. (Ph.D.). Stanford University.

Arora, Jagdish, and Anju Vyas. 1998. Providing structured access to the electronic journals in engineering and technology on the Internet: An analysis. In *Proceeding of the 49th FID Conference and Congress,* III-90-III-99. The Hague: FID.

Artemeva, Natasha, and Aviva Freedman. 2001. "Just the boys playing on computers." An activity theory analysis of differences in the cultures of two engineering firms. *Journal of Business and Technical Comunication* 15(2): 164–194.

Arthur, Richard H. 1980. Developing a broad based communication course for engineers. *66th Annual Meeting of the Speech Communication Association,* New York, November 13–16, 1980. (Available from ERIC ED 197392.)

Arvai, Joseph L. 2000. Evaluating NASAs role in risk communication process surrounding space policy decisions. *Space Policy* 16: 61–69.

Association of American Universities Research Libraries Project in collaboration with the Association of Research Libraries. 1994. *Reports of the AAU Task Forces: On Acquisition and Distribution of Foreign Language and Area Studies Materials—A National Strategy for Managing Scientific and Technological Information: Intellectual Property Rights in an Electronic Environment.* Washington, DC: Association of Research Libraries.

Athanassiades, J. C. 1973. The distortion of upward communication in hierarchical organizations. *Academy of Management Journal* 16(2): 207–226.

Atkinson, Roderick D., and Laurie E. Stackpole. 1995. TORPEDO: Networked access to full-text and page-image representations of physics journals and technical reports. *Public-Access Computer Systems Review* 6(3): 6–15. Available from World Wide Web: <http://info.lib.uh.edu/pr/v6/n3/atki6n3.html>.

Atman, Cynthia J., Karen M. Bursic, and Stefanie L. Lozito. 1995. Gathering information: What do students do? In *1995 ASEE Conference Proceedings,* pp. 1138–1144. Washington, DC: ASEE.

Auerbach Corporation and Arnold F. Goodman, et al. 1965. *DOD User Needs Study: Phase 2.* North American Aviation. Available from NTIS AD616501, AD616502, AD647111, AD647112, AD649284.

Augustine, N. R., and C. M. Vest. 1994. *Engineering Education for a Changing World.* Washington: ASEE.

Auster, E. 1982. Organizational behavior and information seeking: lessons for librarians. *Special Libraries* 73(3): 173–182.

Auster, Ellen R. 1990. The Interorganizational Environment: Network theory, tools and applications. In *Technology Transfer: A Communication Perspective.* Frederick Williams and David V. Gibson, eds., pp. 63–89, Newbury Park, CA: Sage Publications.

Badawy, M. K. 1982. *Developing Managerial Skills in Engineers and Scientists.* New York: Van Nostrand Reinhold.

Baird, F., C. J. Moore, and A. P. Jagodzinski. 2000. An ethnographic study of engineering design teams at Rolls-Royce Aerospace. *Design Studies* 21: 333–355.

Baker, D. B. 1970. Communication or chaos. *Science* 169: 739–742.

Baker N. L., et al. 1980. Idea generation: A procrustean bed of variables, hypotheses, and implications. In: B. V. Dean and J. L. Goldhar, eds., *Management of Research and Innovation,* pp. 33–51. Amsterdam: North-Holland.

Bakos, J. D., Jr. 1997. Communication skills for the 21st century. *Journal of Professional Issues in Engineering Education and Practice* 123(1): 14–16.

Baldwin, John, and David Sabourin. 2000. Innovative activity in Canadian food processing establishments: The importance of engineering practices. *International Journal of Technology Management* 20(5/6/7/8): 511–527.

Baltatu, Monica E. 1984. Online Information. *Chemical Engineering* (January): 69–72.

Barchilon, M. G. 1993. Topic integration: How technical communication courses can equip engineers for industry. In *23rd Annual Frontiers in Education Conference,* pp. 384–386. IEEE.

Barchilon, Marian G., and Robert Baren. 1998. Improving engineering communication through the long-distance classroom. In *Proceedings of the 1998 28th Annual Frontiers in Education Conference,* pp. 999–1002. Piscataway, NJ: IEEE.

Barclay, Rebecca O., Thomas E. Pinelli, and John M. Kennedy. 1993. A comparison of the technical communication practices of Dutch and U.S. aerospace engineers and scientists. Report 17. Washington, DC: National Aeronautics and Space Adminstration. NASA TM-108987. (Available from NTIS 94N11352.)

Barclay, Rebecca O., Thomas E. Pinelli, and John M. Kennedy. 1994. Technical communication practices of Dutch and U.S. aerospace engineers and scientists: International perspectives on aerospace, paper 41. Reprinted from *IEEE Transactions on Professional Communication* 37:2 (June), pp. 97–107. (Available from AIAA.)

Barclay, Rebecca O., Thomas E. Pinelli, David Elazar, and John M. Kennedy. 1991. An analysis of the technical communications practices reported by Israeli and U.S. aerospace engineers and scientists, paper 14. Paper presented at the *International Professional Communication Conference (IPCC).* Available from NTIS 92N28183.

Barclay, Rebecca O., Thomas E. Pinelli, Michael L. Keene, John M. Kennedy, and Myron Glassman. 1991. Technical communications in the international workplace: Some implications for curriculum development, paper 15. Reprinted from *Technical Communication,* 38, 3 (3rd Quarter, August), pp. 324–335. (Available from NTIS 92N28116.)

Barclay, Rebecca O., Thomas E. Pinelli, Axel S. T. Tan, and John M. Kennedy. 1993. Technical communications practices and the use of information technologies as reported by Dutch and U.S. aerospace engineers. In *Proceedings of the 1993 IEEE International Professional Communications Conference,* pp. 221–226. Piscataway, NJ: IEEE.

Barczak, Gloria, and David Wilemon. 1991. Communication patterns of new product development team leaders. *IEEE Transactions on Engineering Management* 38(2): 101–109.

Baren, M. Robert, and James Watson. 1993. Developing communication skills in engineering classes. In *Proceedings of the 1993 IEEE International Professional Communication Conference,* pp. 432–437. Piscataway, NJ: IEEE.

Barnard, C. I. 1946. *The Functions of the Executive.* Boston: Harvard University Press.

Barnes, R. C. M. 1965. Information use studies. Part 2—Comparison of some recent surveys. *Journal of Documentation* 21(3): 169–176.

Barnett, A. A., and J. E. Russell. 1994. Development of a petroleum engineering electronic journal. In *Proceedings of the 1994 Petroleum Computer Conference,* pp. 129–134. Richardson, TX: SPE.

Barreau, Deborah K. 1995. Context as a factor in personal information management systems. *Journal of the American Society for Information Science* 46: 327–339.

Barreau, Deborah K., and Bonnie A. Nardi. 1995. Finding and reminding: File organization from the desktop. *ACM SIGCHI Bulletin* 27: 39–43.

Bartenbach, Bill. 1996. Organizing the engineering resources on the Internet. *IATUL Proceedings* 5. Available from World Wide Web: <http://educate. lib.chalmers.se/IATUL/ proceedcontents/abs196/Bartenba.html>.

Baskaran, A. 2001. Competence building in complex systems in the developing countries: The case of satellite building in India. *Technovation* 21(2): 109–121.

Baskin, O. W., and C. E. Aronoff. 1980. *Interpersonal Communication in Organizations.* CA: Goodyear Publishing Co.

Bates, Marcia J. 1996. Learning about the information seeking of interdisciplinary scholars and students. *Library Trends* 45(2): 155–164.

Batson, Robert G. 1987. Characteristics of R&D management which influence information needs. *IEEE Transactions on Engineering Management* EM-34, no. 3 (August): 178–83.

Battelle Columbus Laboratories. 1965. Interactions of science and technology in the innovative process: Some case studies. Washington, DC: National Aeronautics and Space Administration. (Available NTIS PB228508.)

Batts, Martin. 1995. Opening of the engineering mind. In *1995 ASEE Conference Proceedings,* pp. 413–415. Washington, DC: ASEE.

Bavelas, A. 1968. Communication patterns in task-oriented groups. In *Group Dynamics: Research and Theory,* 3rd ed. D. Cartwritght and A. Zander, eds. pp. 503–511. London: Tavistock Publishers.

Bayer, Alan E., and Gerald Jahoda. 1981. Effects of online bibliographic searching on scientists' information style. *Online Review* 5(4): 323–333.

Bayer, Alan E., and Gerald Jahoda. 1979. Background characteristics of industrial and academic users and nonusers of online bibliographic search services. *Online Review* 3(1): 95–105.

Beardsley, Charles W. 1972. Keeping on top of your field. *IEEE Spectrum* (December): 68–71.

Beck, Charles E. 1994. Creating a climate for teamwork. In *Proceedings of the 1994 IEEE International Professional Communications Conference,* pp. 222–226. Piscataway, NJ: IEEE.

Becker, L. G. 1980. Information resources management (IRM): A revolution in progress. *ASIS Bulletin* 6(6): 26–27.

Beckert, Beverly A. 1988. The technical office. *Computer-Aided Engineering: CAE* 7(12): 76–80.

Beer, D. F., and D. A. McMurrey. 1997. *A Guide to Writing as an Engineer.* New York: Wiley.

Belefant-Miller, Helen, and Donald W. King. 2001. How, what, and why science faculty read. *Science & Technology Libraries* 19(2): 91–107.

Belew, W. W. 1980. *Passive Solar Energy Information User Study.* Golden, CO: Solar Energy Institute.

Bender, Laura J., Robert C. Chang, Patricia Morris, and Chris Sugnet. 1997. A science-engineering library's needs assessment survey: Method and learnings. *Science & Technology Libraries* 17(1): 19–34.

Bennis, W. G. 1961. Interpersonal communication. In *The Planning of Change*. W. G. Bennis et al., eds. New York: Holt, Reinhart and Winston.

Bensahel, J. G. 1981. Mixing the formal with the informal. *International Management* 36(1): 37–38.

Berge, Z. L., and M. P. Collins. 1996. IPCT journal readership survey. *Journal of the American Society for Information Science* 47(9): 701–10.

Berger, C. R., and M. E. Roloff. 1980. Social cognition, self-awareness, and interpersonal communication. In *Progress in Communication Sciences*. B. Dervin and M. J. Voigt, eds., 2: 1–49.

Berkowitz, N. H., and W. G. Bennis. 1961/62. Interaction patterns in formal service-oriented organizations. *Administrative Science Quarterly* 6: 25–50.

Bermar, Amy. 1990. Babel in Design Land: "Can we talk?" Marketing and design don't speak the same language; In fact, they often don't even speak. *EDN* 35(18A): 57–58.

Berry, Frederick C., and Patricia A. Carlson. 1999. Asynchronous learning environment for integrating technical communication into engineering courses. In *Proceedings of the 29th ASEE/IEEE Frontiers in Education Conference,* 13a6-21. Piscataway, NJ: IEEE.

Berul, Lawrence H., et al. 1965. *DOD User Needs Study: Phase 1*. Philadelphia, PA: Auerbach Corporation. Available from NTIS AD616501, AD616502, AD649284.

Bhalla, S. K. 1981. Make R&D cheaper, faster, better with teamwork. *Industrial Research and Development* 23(9): 159–161.

Bhattacharya, S. C. 2001. Renewable energy education at the university level. *Renewable Energy* 22(1): 91–97.

Bichteler, Julie. 1991. Geologists and gray literature: Access, use, and problems. *Science and Technology Libraries* 11(3): 39–49.

Bichteler, Julie, and Dederick Ward. 1989. Information-seeking behavior of geoscientists. *Special Libraries* 79(3): 169–178.

Bikson, T. K., B. E. Quint, and L. L. Johnson. 1984. *Scientific and Technical Information Transfer: Issues and Options*. Santa Monica, CA: Rand Corp.

Bishop, Ann P. 1994. The role of computer networks in aerospace engineeing, paper 39. Paper presented at the *32nd Aerospace Sciences Meeting of the American Institute of Aeronautics and Astronautics (AIAA), Reno, NV, January 1994*. (Available from AIAA 94-0841.) Also available from *Library Trends* 42(4): 694–729.

Bishop, A., and M. O. Fellows. 1989. Descriptive analysis of major federal scientific and technical information policy studies. In *U.S. Scientific and Technical Information (STI) Policies: Views and Perspectives*, C. R. McClure and P. Hernon, eds. Norwood, NJ: Ablex.

Bishop, Ethelyn, and Audry Clayton. 1976. *User Values of Information Service Characteristics*. Arlington, VA: Forecasting International, Ltd.

Blade, Mary Frances. 1963. Creativity in engineering. In *Essays on Creativity in the Sciences*. New York: New York University Press.

Blados, Walter R., Thomas E. Pinelli, John M. Kennedy, and Rebecca O. Barclay. 1990. External information sources and aerospace R&D: The use and importance of technical reports by U.S. aerospace engineers and scientists. paper 2. Paper prepared for the *68th AGARD National Delegates Board Meeting,* March 29, 1990, Toulouse, France. (Available from NTIS 90N30132.)

Blagden, John, John Harrington, and Heather Woodfield. 1994. *Eurilia (European Initiative in Library and Information in Aerospace): An Audit of Aerospace Information Needs in Five European Countries,* College of Aeronautics Report No. 9405. Bedford: Cranfield University.

Blagden, John, John Harrington, and Heather Woodfield. 1997. *Aerospace Information Needs in Europe: Results of the Post Audit Study,* College of Aeronautics Report No. 9705. Cranfield: Cranfield University.

Blake, Gary. 1998. Watching what you write. *IIE Solutions* 30(1): 38–39.

Bland, M. 1980. *Employee Communications in the 1980s.* London: Kogan Page Publishers.

Blaxter, K. L., and M. L. Blaxter. 1973. The individual and information problem. *Nature* 246: 335–339.

Bommer, M., and D. Jalajas. 2002. The innovation work environment of high-tech SMEs in the USA and Canada. *Research & Development Management* 32(5): 379–386.

Boot, J. C. G., and E. B. Cox. 1970. *Statistical Analysis for Mangerial Decisions.* New York: McGraw-Hill.

Borchardt, John K. 1990. Improve in-house communications. *Chemical Engineering* 97(3): 135–138.

Borgman, Christine L. 1989. All users of information retrieval systems are not created equal: An exploration into individual differences. *Information Processing and Management* 25(3): 237–251.

Borgman, Christine L., Donald O. Case, and Charles T. Meadow. 1985. Incorporating users' information seeking styles into the design of an information retrieval interface. In *Proceedings of the 48th ASIS Annual Meeting* 22: 324–330.

Borgman, Christine L., Donald O. Case, and Charles T. Meadow. 1986. Evaluation of a system to provide online instruction and assistance in the use of energy databases: The DOE/OAK project. In *Proceedings of the 49th ASIS Annual Meeting* 23: 32–38.

Borgman, Christine L., Donald O. Case, and Charles T. Meadow. 1989. The design and evaluation of a front-end user interface for energy researchers. *Journal of the American Society for Information Science* 40(2): 99–109.

Borman, Stu. 1993. Advances in electronic publishing herald changes for scientists. *Chemical & Engineering News* 71(24): 10–24.

Bormann, E. G., et al. 1969. *Interpersonal Communication in the Modern Organization.* Englewood Cliffs, NJ: Prentice-Hall.

Borvovansky, Vladimir T. 1996. Changing trends in scholarly communication: Issues for technological university libraries. In *IATUL Proceedings.* Available from World Wide Web: <http://educate.lib.chalmers.se/IATUL/proceedcontents/abs196/Borovan.html>.

Boston, O. P., Stephen J. Culley, and Christopher A. McMahon. 1996. Designers and suppliers modelling the flow of information. In *Proceedings of ILCE '96.* Paris.

Bouazza, Abdelmajid. 1989. Information user studies. In *Encyclopedia of Library and Information Science* 44, Allen Kent, ed., pp. 144–164. New York: Marcel Dekker.

Boulgarides, J. D., and V. C. San Filippo. 1969. How do engineers look at continuing education? *Professional Engineer* 39(3): 32–37.

Brady, Edward L., ed. 1985. U.S. access to Japanese technical literature: Electronics and electrical engineering (NBS Special Publication 710). Proceedings of a seminar held at the National Bureau of Standards, Gaithersburg, Maryland, USA, June 24–25, 1985. Washington, DC: U.S. Department of Commerce, National Bureau of Standards, 1986.

Braham, James. 1991. Captains of video: Through the marvels of videoconferencing, engineers are slicing development time as well as travel expense. *Machine Design* 63(9): 71–75.

Brasey, E. 1983. L'anglais ne suffit pas. *L'Expansion* 208: 21.

Branin, Joseph J., and Mary Case. 1998. Reforming scholarly publishing in the sciences: A librarian perspective. *Notices of the AMS* 45(4): 475–486.

Branscomb, Lewis M. 1992. U.S. scientific policy and technical information policy in the context of a diffusion-oriented national technology policy. *Government Publications Review* 19: 189–193.

Breitzman, Anthony F. 2003. *An Objective Analysis of the Effect of IEEE Publications on Subsequent Patented Technology.* Piscataway, NJ: IEEE. Available from the World Wide Web <*www.ieee.org/patentcitation*>.

Breton, E. J. 1981. Why engineers don't use databases. *ASIS Bulletin* 7(6): 20–23.

Breton, E. J. 1981. Reinventing the wheel: The failure to use existing technology. *Mechanical Engineering* 103(3): 54–57.

Brinberg, Herbert R., and Thomas E. Pinelli. 1993. A general approach to measuring the value of aerospace information products and services, paper 24. Paper presented at the *31st Aerospace Sciences Meeting and Exhibits of the American Institute of Aeronautics and Aerospace (AIAA),* Reno, NV, January 11–13, 1993. AIAA-93-0580. (Available from AIAA 93A17511.)

Brittain, J. M. 1970. *Information and Its Uses.* New York: Wiley-Interscience, pp. 146–152.

Broadbent, Marianne, and Hans Lofgren. 1991. *Priorities, Performance and Benefits: An Exploratory Study of Library and Information Units.* Melbourne: Centre for International Research on Communication and Information Services.

Brooke, A. 1980. Some user education techniques appropriate to special libraries. *Australian Special Libraries News* 13(1): 15–18.

Brookes, B. C. 1963. Communication between scientists. *Advancement of Science* 19: 559–563.

Brookes, N. J., C. J. Backhouse, and N. D. Burns. 1994. Modeling product introduction: A key step to the successful concurrent engineering applications. In *Proceedings of the Tenth National Conference on Manufacturing Research: Advances in Manufacturing Technology VIII,* pp. 452–456. Loughborough, UK: Loughborough University of Technology.

Brown, Cecilia M. 1999. Information seeking behavior of scientists in the electronic information age: Astronomers, chemists, mathematicians, and physicists. *Journal of the American Society for Information Science* 50(10): 929–943.

Brown, James William. 1979. The technological gatekeeper: Evidence in three industries. *Journal of Technology Transfer* 3(2): 23–36.

Brown, W. B. 1969. Systems, boundaries and information flow. In *Readings in Management,* 3rd ed. M. D. Richards and W. A. Nielander, eds., pp. 121–133. Mason, OH: South-Western Publishing Company.

Buckley, Chad, and Amy Prendergast. 1998. Wired research: The accessibility of electronic journals in science and technology. *Issues in Sciences and Technology Librarianship* 17. Available from World Wide Web: <http://www.library.ucsb.edu/istl/98-winter/conference5.html>.

Buntrock, Robert E., and Aldona K. Valicenti. 1985. End-users and chemical information. *Journal of Chemical Information and Computer Science* 25: 203–207.

Burdan, Amy L., and Judith B. Strother. 1995. Are your communication skills good enough. In *Proceedings of the 1995 IEEE International Professional Communication Conference,* pp. 156–160. Piscataway, NJ: IEEE.

Burk, Cornelius F., and Forest W. Horton. 1988. *Info Map; A Complete Guide to Discovering Corporate Information Resources.* Englewood Cliffs, NJ: Prentice-Hall.

Burnett, Kathleen, and E. Graham McKinley. 1998. Modelling information seeking. *Interacting with Computers* 10: 285–302.

Burns, T. 1954. The directions of activity and communication in a departmental executive group. *Human Relations* 7: 73–97.

Burte, Harris M. 1975. Some experiences in generating and maintaining communication within interdisciplinary R&D teams. *Joint Engineering Management Conference,* October 9–10, 1975, pp. 71–73.

Button, Graham, and Wes Sharrock. 1996. Project work: The organization of collaborative design and development in software engineering. *Computer Supported Cooperative Work: The Journal of Collaborative Computing* 5: 369–386.

Byström, Katrina, and Kalervo Järvelin. 1995. Task complexity affects information seeking and use. *Information Processing & Management* 31(2): 191–213.

Campbell, D. T. 1958. Systematic error on the part of human links in communication systems. *Information and Control* 1: 334–369.

"Canada: Saskatoon Engineers Get Serious." *Engineering Digest* 33 (April 1987): 4.

Carlson, Patricia A. 1999. Cognitive foundations for teaching technical communication. In *Proceedings of the 29th ASEE/IEEE Frontiers in Education Conference,* 13a6-20. Piscataway, NJ: IEEE.

Carstensen, Peter H. 1997. Towards information exploration support for engineering designers. In *Advances in Concurrent Engineering—CE97,* pp. 26–33. Lancaster, PA: Technomic.

Carter, M. P. 1983. Determining information needs. *Management Decisions* 21(4): 45–51.

Carter R. M. 1972. *Communication in Organizations: An Annotated Bibliography.* Farmington Hills, MI: Gale Research Co.

Cartwright, D., and A. Zander. 1968. *Group Dynamics: Research and Theory,* 2nd ed. London: Tavistock.

Case, Donald, Christine Borgman, and Charles Meadow. 1986. End-user information-seeking in the energy field: Implications for end-user access to DOE/RECON databases. *Information Processing & Management* 22(4): 299–308.

Case, Donald, Christine Borgman, and Charles Meadow. 1985. Information-seeking in the energy research field: The DOE/OAK project. In *Proceedings of the 48th ASIS Annual Meeting* 22: 331–336.

Case, Mary M. 1998. Recreating publishing for the engineering and scientific community: The Scholarly Publishing & Academic Resources Coalition (SPARC). In *Proceedings of the Socioeconomic Dimensions of Electronic Publishing Workshop,* pp. 71–75. Piscataway, NJ: IEEE.

Castelo, V. 1993. Bulletin boards, electronic mail, conferencing, current use by scientists and engineers: Effects on libraries and information centres: Do they have a role? In *AGARD, International High Speed Networks for Scientific and Technical Information.*

Castelyn, M. 1981. *Planning Library Training Programmes.* London: Andre Deutsch.

Castillo, Alicia. 2000. Ecological information system: Analyzing the communication and utilization of scientific information in Mexico. *Environmental Management* 25(4): 383–392.

Cerri, Steven. 1999. Effective communication skills for engineers. *IEEE Antennas and Propagation Magazine* 41(3): 100–103.

Chakrabarti, A. K., et al. 1983. Characteristics of sources, channels and contents for scientific and technical information systems in industrial R&D. *IEEE Transcripts on Engineering Management* EM-30(2): 83–88.

Chakrabarti, A. K., and A. H. Rubenstein. 1976. Interorganizational transfer of technology: A study of adoption of NASA innovations. *IEEE Transactions on Engineering Management.* EM-23, no. 1 (February): 20–34.

Chang, Peter C., Richard H. McCuen, and Jayanta K. Sircar. 1995. Multimedia-based instruction in engineering education: Strategy. *Journal of Professional Issues in Engineering Education and Practice* 121(4): 216–219.

Chang, Shan-Ju. 1989. *Information Uses: Relating Information Needs to Information Uses in Specific Contexts.* Available from ERIC 348 999.

Chapanis, A. 1975. Interactive human communication. *Scientific American* 232(3): 36–42.

Chen, Ching-Chih, and P. Hernon. 1982. *Information Seeking.* New York: Neil-Schuman Publishers.

Cherry, C. 1978. *On Human Communication,* 3rd ed. Boston: MIT.

Cheuk, Wai-Yi Bonnie. 1998. Exploring information literacy in the workplace: A qualitative study of engineers using the sense-making approach. *International Forum on Information and Documentation* 23(2): 30–38.

Chinowsky, Paul S., and Meghann A. Byrd. 2001. Strategic management in design firms. *Journal of Professional Issues in Engineering Education and Practice* 127(1): 32–40.

Christie, B. 1981. *Face to File Communication: A Psychological Approach to Information Systems.* London: Wiley.

Clark, D. F., and R. L. Ackoff. 1959. A report on some organizational experiments. *Operations Research* 7: 279–293.

Clayton, Audrey. 1981. The potential influence of social, economic, regulatory and technological factors on scientific and technological communication through 2000 AD. Final Report. Washington, DC: Division of Information Science and Technology, National Science Foundation.

Clutterbuck, D. 1980. Breaking through the cultural barrier. *International Management* 35(12): 41–42.

Cohen, A. M. 1961/62. Changing small-group communication networks. *Administrative Science Quarterly* 6: 443–462.

Coleman, J. S. 1958. Relational analysis: The study of social organizations with survey methods. *Human Organizations* 17: 28–36.

Coleman, J. S., and D. MacRea. 1960. Electronic processing of sociometric data for groups up to 1000 in size. *American Sociological Review* 25(5): 722–727.

Collins, H. M. 1974. The TEA set: Tacit knowledge and scientific networks. *Science Studies* 4(2): 165–186.

Collins, Robyn, Shuyun Li, and Dorothy Cheung. 2000. Language professionals in engineering faculty: Cross-cultural experience. *Journal of Professional Issues in Engineering Education and Practice* 126(1): 32–34.

Commission of the European Communities. 1968/1970. Les cadres juridiques de la coopération internationale en matière scientifique et le problème europèen. *Actes de Colloques d'Aix-en-Provence, Dec. 1–2, et Nice, France, 1968.* CEC, Dec. 6–7, 1970.

Condous, C. 1983. Non-profit marketing: Libraries' future? *ASLIB Proceedings* 35(10): 407–417.

Constance, J. D. 1982. Creating a professional work climate. *Machine Design* 54(19): 94–98.

Correia, A. M. R. 1993. Scientific and technical information (STI) towards technological and industrial development: The case of Portugal. *Journal of Information Science* 19(1): 25–35.

Corridore, Michael C. 1976. Scientific and technical information needs and users or potential users of the DSA-administered DOD information analysis centers: Final report. (NTIS AD-A024 937). Alexandria, VA: DSA.

Council on Library Resources, Inc. 1990. *Communications in Support of Science and Engineering: A Report to the National Science Foundation from the Council on Library Resources.* Washington, DC: Council on Library Resources.

Court, Andrew W. 1995. The classification and modeling of information for engineering designers. (Ph.D. thesis). University of Bath.

Court, Andrew W., Stephen J. Culley, and Christopher A. McMahon. 1997. The influence of information technology in new product development: Observations of an empirical study of the access of engineering design information. *International Journal of Information Management* 17(5): 359–375.

Court, Andrew W., Stephen J. Culley, and Christopher A. McMahon. 1996. Information access diagrams: A technique for analyzing the usage of design information. *Journal of Engineering Design* 7(1): 55–75.

Court, Andrew W., Stephen J. Culley, and Christopher A. McMahon. 1995. Modeling the information access methods of engineering designers. In *Proceedings of the 1995 ASME Design Engineering Technical Conference,* pp. 547–554. New York: ASME.

Court, Andrew W., Stephen J. Culley, and Christopher A. McMahon. 1994a. The importance of information transfer between the designer and manufacturer. In *Proceedings of the 4th International Conference on Flexible Automation and Information Management,* pp. 352–362. Blacksburg, VA.

Court, Andrew W., Stephen J. Culley, and Christopher A. McMahon. 1994b. Information sources and storage methods for engineering data. In *Proceedings of the 2nd ASME Biennial European Joint Conference on Engineering Systems Design & Analysis,* PD Vol. 64–5, 9–16. London.

Coyne, J. G., T. E. Hughes, and B. C. Winsbro. 1986. Sharing results of federal R&D: A look at the department of energy's system for managing scientific and technical information. *Government Information Quarterly* 3: 363–380.

Crane, Diana. 1972. *Invisible Colleges: Diffusion of Knowledge in Scientific Communities.* Chicago: University of Chicago Press.

Crane, Diana. 1971a. Information needs and users. In *Annual Review of Information Science and Technology* 6: 3–39. Chicago: Encyclopaedia Brittannica.

Crane, Diana. 1971b. Transnational networks in basic science. *International Organization* 25(3): 585–601.

Crane, Diana. 1970. The nature of scientific communication and influence. *International Social Science Journal* 22(1): 28–41.

Crane, Diana. 1969. Social structure in a group of scientists: A test of the "invisible college" hypothesis. *American Sociological Review* 34(3): 335–352.

Crane, Diana. 1967. The gatekeepers of science: Some factors affecting the selection of articles for scientific journals. *American Sociologist* 2(4): 195–201.

Crawford, Susan. 1978. Information needs and uses. In *Annual Review of Information Science and Technology* 13: 61–81. White Plains: Knowledge Industry Publications.

Crawford, Susan. 1971. Informal communication among scientists in sleep research. *JASIS* 22(5): 301–310.

Cronin, B. 1981. The marketing of library and information services. *ASLIB.*

Culley, Stephen J., and Pravir K. Chawdry. 1996. Flying around the wide world of the Web. *Design Engineering* (January): 24–26.

Culley, Stephen J., and A. P. Wallace. 1994. The modelling of engineering assemblies based on standard components. In *Computer Aided Conceptual Design, Lancaster Workshop on Engineering Design,* pp. 113–130. Lancaster, UK.

Culnan, M. J., and J. H. Bair. 1983. Human communication needs and organizational productivity: The potential impact of office automation. *JASIS* 34(3): 215–221.

Cumming, Denise. 1981. But I need it today: Information transfer in an applied research environment. *Proceedings of the Technology Transfer Society International Symposium.* Indianapolis, IN: Society Meeting June 14–17, 1981. Washington, DC, pp. 6.21–6.22.

Curl, Sheila R. 2001. Subramanyam revisited: Creating a new model for information literacy instruction. *College & Research Libraries* 62(5): 455–464.

Daft, R. L., and R. H. Lengel. 1983. *Information Richness: A New Approach to Managerial Behavior and Organization Design.* College Station, TX: Texas A&M University.

Dalton, Marie, and Charles Dalton. 1975. Body talk: A discussion of nonverbal communication in engineering groups. *Joint Engineering Management Conference* (October 9–10, 1975): 47–51.

Davey-Wilson, Ian E. G. 2001. Innovation in the building process: A postgraduate module. *Engineering Structures* 23(1): 136–144.

Davies, A. H. 1976. *The Practise of Marketing Research.* London: Heinemann.

Davies, J. 1983. Linguistic and political barriers in the international transfer of information in science and technology: A re-interpretation. *Journal of Information Science* 6(5): 179–181.

Davis, K. 1981. *Human Behavior at Work: Organizational Behavior.* New York: McGraw-Hill.

Davis, K. 1969. Communication within management. In *Readings in Management,* 3rd ed. M. D. Richards and W. A. Nielander, eds., pp. 155–162. Mason, OH: South-Western Publishing Co.

Davis, K. 1953a. A method of studying communication patterns in organizations. *Personal Psychology* 6: 301–312.

Davis, K. 1953b. Management communication and the grapevine. *Harvard Business Review* 31: 43–49.

Davis, Peter, and Marcia Wikof. 1988. Scientific and technical information transfer for high technology: keeping the figure in its ground. *R&D Management* 18(1): 45–58.

Davis, Richard M. 1975. Technical writing: Its place in engineering curricula—a survey of the experience and opinions of prominent engineers. Wright-Patterson Air Force Base, OH. (Available from NTIS ADA015906.)

Dedert, Patricia L., and David K. Johnson. 1989. Promoting and supporting end-user online searching in an industrial research environment: A survey of experiences at Exxon Research and Engineering Company. *Science and Technology Libraries* 10(1): 25–45.

DeFanti, Thomas A., and Maxine D. Brown. 1989. Scientific animation workstations: Creating an environment for remote research, education, and communication. *Academic Computing* 3(6): 10–12, 56–57.

Denning, Peter J., and Bernard Rous. 1994. *The ACM electronic publishing plan.* New York: ACM.

Dennison, E., et al. 1983. International aid: Channels of communication. *Management Decisions* 21(4): 36–44.

Derr, Richard L. 1982. A conceptual analysis of information need. *Information Processing & Management* 19(5): 273–278.

Dervin, Brenda, and Michael Nilan. 1986. Information needs and uses. In *ARIST* 21: 3–33. White Plains, NY: Knowledge Industry Publications.

Dervin, Brenda. 1980. Communication gaps and inequities: Moving towards a re-conceptualization. In *Progress in Communication Sciences*. B. Dervin and M. J. Voigt, eds., 2: 73–112.

Dessouky, Ibtesam A. 1996. Navigating the Internet for engineering information. *Computers & Industrial Engineering* 31(3–4): 889–92.

Dessouky, Ibtesam A. 1994. The relation between the extended and main libraries for engineering education. *Illinois Libraries* 76: 49–52.

Deutsch, M. 1968. The effects of cooperation and competition upon group process. In *Group Dynamics: Research and Theory,* 3rd ed. D. Cartwright and A. Zander, eds., pp. 461–482. London: Tavistock Publishers.

Dewhirst, H. D., et al. 1978. Satisfaction and performance in research and development tasks as related to information accessibility. *IEEE Transcripts on Engineering Management* EM-25(3): 58–63.

Dewhirst, H. D. 1970. *The Role of the LRC Technical Library in Fulfilling the Information Needs of Professional Employees.* Langley, VA: Langley Research Center.

Dimitrakis, William J. 1997. Communicating with Asian engineers. *Power Transmission Design* 39(8): 56–58.

Disch, A. 1976. The voice of the user: His information needs and requirements. In *The Problem of Optimization of User Benefit in Scientific and Technological Information Transfer*. AGARD.

Dixon, John R. 1991. Information infusion is strategic management. *Information Strategy: The Executive's Journal* 8(1): 16–21.

Dlaska, A. 1999. Suggestions for a subject-specific approach in teaching foreign languages to engineering and science students. *System* 27: 401–417.

Dobrowolski, M. 1981. Information for engineers: A preference for informal channels. *Instrument Technology* 28(6): 14–16.

Dodd, S. C., and M. McCurtain, 1965. The linguistic diffusion of information through randomly overlapped cliques. *Operational Research Quarterly* 16(1): 51–63.

Dordick, H. S., et al. 1981. *The Emerging Network Marketplace.* Greenwich, CT: Ablex Publishing.

Downs, A. 1967. *Inside Bureaucracy.* London: Little, Brown.

Doty, Philip. 1992. Electronic networks and social change in science. *Proceedings of the 55th Annual Meeting of the American Society for Information Science* 29: 185–192.

Drucker, Barry J. 1993. Share your experiences with others through publications. *Journal of Environmental Health* 55(June): 40–41.

Dunn, O. J., and V. A. Clark. 1974. *Applied Statistics: Analysis of Variance and Regression.* London: Wiley.

Dunning, A. 1983. The INSIS programme: Information technology for the office worker. In *7th Annual Online Information Meeting.*

Dunning, A. 1982. INSIS: Inter-institutional integrated services information system. In *Proceedings of 6th International Online Information Meeting.*

Dyke, Julie, and Patricia Wojahn. 2000. Getting "dissed": Technical communicators in interdisciplinary engineering teams. In *Technology & Teamwork,* pp. 7–23. IEEE.

Dykeman, Amy. 1994. Faculty citations: An approach to assessing the impact of diminishing resources on scientific research. *Library Acquisitions: Practice and Theory* 18(2): 137–146.

Eason, Ken, Chris Carter, Susan Harker, Sue Pomfrett, Kathy Phillips, and John Richardson. 1997. *A Comparative Analysis of the Role of Multi-media Electronic Journals in Scholarly Disciplines.* Loughborough: HUSAT Research Institute and Department of Human Sciences, Loughborough University. Available from World Wide Web: <http://www.ukoln. ac.uk/services/elib/papers/tavistock/eason/eason.html>.

Eckerson, Wayne. 1990. DEC's net makes the world one big office. *Network World* 7(27): 17–18.

Egan, F. T. 1982. Today's engineer: A survey. *Electronic Products* 25(9): 40–45.

Ellis, D., D. Cox, and K. Hall. 1993. A comparison of the information seeking patterns of researchers in the physical and social sciences. *Journal of Documentation* 49(4): 356–369.

Ellis, David, and Merete Haugan. 1997. Modelling the information seeking patterns of engineers and research scientists in an industrial environment. *Journal of Documentation* 53(4): 384–403.

Ellis, Richard A. 1982. Information acquisition and processing by the technical person: Seance, science, synthesis. In *11th ASIS Midyear Meeting,* Knoxville, TN, June 13–16, 1982.

Emery, F. E. 1971. *Systems Thinking: Selected Readings.* New York: Penguin.

Emrich, B. R. 1970. *Scientific and Technical Information Explosion.* NTIS.

Engineering Index, Inc. 1978. *Alternatives for Accessing Engineering Numerical Data.* NTIS.

Enslein, K., et al. 1977. *Statistical Methods for Digital Computers.* London: Wiley.

Er, M. C. 1989. A critical review of the literature on the organizational impact of information technology. *IEEE Technology and Society Magazine* (June): 17–23.

Ercegovac, Zorana. 1988. *Variables Considered in the Design of an Environmental Information Center.* Los Angeles: UCLA Engineering Research Center.

Ercegovac, Zorana. 1999. Learning portfolio for accessing engineering information for engineers. In *Proceedings of the ASIS Annual Meeting,* pp. 450–461. Silver Spring, MD: ASIS.

Esler, Sandra L., and Michael L. Nelson. 1998. Evolution of scientific and technical information distribution. *Journal of the American Society for Information Science* 49(1): 82–91.

Estabrook, Leigh Stewart. 1986. Valuing a document delivery system. *RQ* 26(1): 58–62.

Etnier, Carl. 1986. Secrecy and the young researcher. *Technology in Society* 8: 276–271.

Ettlie, John E. 1976. The timing and sources of information for the adoption and implementation of production innovations. *IEEE Transactions on Engineering Management* 23(1): 62–68.

Evans, A. J., et al. 1977. *Education and Training of Users of Scientific and Technical Information: UNISIST Guide for Teachers.* UNESCO.

Eveland, J. D., and Thomas E. Pinelli. 1991. Information intermediaries and the transfer of aerospace scientific and technical information (STI): A report from the field. paper 9. Paper commissioned for presentation at the *1991 NASA STI Annual Conference,* April 9, 1991, Huntsville, AL. (Available from NTIS 91N21959.)

Exon, A. 1978. Getting to know the user better. *ASLIB Proceedings* 30(10/11): 352–364.

Faibisoff, Sylvia G., and Donald P. Ely. 1976. Information and information needs. *Information Reports and Bibliographies* 5(5): 2–16.

Fairbanks, Aline M. 1984. Forensic engineering information services. *Library Trends* (Winter): 303–314.

Farace, R. V., and T. Mabee. 1980. Communication network analysis methods. In *Multivariate Techniques in Human Communication Research.* P. R. Monge and J. N. Cappella, eds. San Diego, CA: Academic Press.

Farace, R. V., et al. 1977. *Communication and Organizing.* Boston: Addison-Wesley.

Farah, Barbara Donald. 1993. The information seeking behavior of academic computing engineers (survey of faculty at eight universities). In *Looking to the Year 2000: 84th Annual Conference, 1993—Special Libraries Association,* pp. 65–102. UMI.

Featheringham, Tom R. 1977. Computerized conferencing and human communication. *IEEE Transactions on Professional Communication* PC-20, Vol. 4 (December): 207–213.

Feibleman, J., and J. W. Friend. 1971. The structure and function of organization. In *Systems Thinking: Selected Readings.* F. E. Emery, ed., pp. 30–50. New York: Penguin.

Feld, W. J., et al. 1983. *International Organizations: A Comparative Approach.* New York: Praeger.

Fenn, M., and G. Head. 1965. Upward communication: The subordinate's viewpoint. *California Management Review* 7(4): 75–80.

Ferguson, S., and S. D. Ferguson. 1980. *Intercom: Readings in Organizational Communciation.* Rochelle Park, NJ: Hayden.

Festinger, L. 1949. The analysis of socioframs using matirx algebra. *Human Relations* 2(3): 153–158.

Festinger, L. 1950. Informal social communcation. *Psychological Review* 57: 271–282.

Festinger, L., and D. Katz. 1953. *Research Methods in the Behavioral Sciences.* London: Dryden.

Fidel, Raya, and Efthimis N. Efthimiadis. 1999. Web searching behavior of aerospace engineers. *SIGIR Forum,* pp. 319–320. New York: ACM.

Fine, Sara. 1984. Research and the psychology of information use. *Library Trends* (Spring): 441–460.

Fischer, W. A. 1980. Scientific and technical information and the performance of R&D groups. In *Management of Research and Innovation.* B. V. Dean and J. L. Goldhar, eds., pp. 67–89. Amsterdam: North-Holland.

Fischer, W. A. 1979. The acquisition of technical information by R&D managers for problem-solveing in non-routine contingency situations. *IEEE Transcripts on Engineering Management* EM-26(1): 8–14.

Fishenden, R. M. 1959. Methods by which research workers find information. In *Proceedings of the International Scientific Conference on Information* 1: 163–179.

Fisher, B. A. 1980. *Small Group Decision-making: Communication and the Group Process,* 2nd ed. New York: McGraw-Hill.

Fisher, Jean, and Susanne Bjorner. 1994. Enabling online end-user searching: An expanding role for librarians. *Special Libraries* 85(4): 281–291.

Fjällbrant, Nancy. 1977. *The Development of a Programme of User Education at Chalmers University of Technology Library* (Ph.D. Thesis).

Fjällbrant, Nancy. 2000. Information literacy for scientists and engineers: Experiences of EDUCATE and DEDICATE. *Program* 34(3): 257–268.

Flammia, Madelyn, Rebecca O. Barclay, Thomas E. Pinelli, Michael L. Keene, Robert H. Burger, and John M. Kennedy. 1993. New era in international technical communication: American-Russian collaboration. In *Proceedings of the 1993 IEEE International Professional Communication Conference,* pp. 217–221. Piscataway, NJ: IEEE.

Flood, B. 1975. Broadcasting vs. narrowcasting: The user as receiver and sender of information. In *Information Revolution. Proceedings of the 38th ASIS Annual Meeting* pp. 45–46.

Flowers, B. H. 1965. Survey of information needs of physicists and chemists. *Journal of Documentation* 21(2): 83–112.

Ford, G. 1973. Research in user behavior. *Journal of Documentation* 20: 85–106.

Franke, Earnest A. 1989. The value of the retrievable technical memorandum system to an engineering company. *IEEE Transactions on Professional Communication* 32(1): 12–16.

Fraser, Emily Jean, and William H. Fisher. 1987. Use of federal government documents by science and engineering faculty. *Government Publications Review* 14: 33–44.

Fraser, Jay. 1992. Plug yourself into a network. *EDN* 37: 221–224.

Freeman, James E., and Albert H. Rubenstein. 1974. The users and uses of scientific and technical information: Critical research needs. Washington, DC: National Science Foundation. (Available NTIS PB237941.)

French, J. R. P. 1968. A formal theory of social power. In *Group Dynamics: Research and Theory,* 3rd ed. D. Cartwright and A. Zander, eds. London: Tavistock.

Fries, James R. 1981. Database searching in chemical engineering. *Chemical Engineering* (December 28, 1981): 71–74.

Frost, Penelope A., and Richard Whitley. 1971. Communication patterns in a research laboratory. *R&D Management* 1(2): 71–79.

Fulltext Sources Online. 2001. Medford, NR: Information Today, Inc.

Gaffney, Inez M. 1976. Users, uses, and supplies of STI services. *Canadian Journal of Information Science* I(1): 35–42.

Galbraith, J. R. 1971. Matrix organization designs. *Business Horizons* pp. 29–40.

Garvey, William D. 1979a. *Communication: The Essence of Science.* New York: Pergamon Press.

Garvey, William D. 1979b. Research studies in patterns of scientific communication, II: The role of the national meeting in scientific and technical communication. In *Communication: The Essence of Science.* W. D. Garvey, pp. 184–201. New York: Pergamon Press.

Garvey, William D. 1979c. Research studies in patterns of scientific communication, III: Information exchange processes associated with the production of journal articles. In *Communication: The Essence of Science.* W. D. Garvey, pp. 202–224. New York: Pergamon Press.

Garvey, William D. 1979d. The dynamic scientific-information user. In *Communication: The Essence of Science.* W. D. Garvey, pp. 256–279. New York: Pergamon Press.

Garvey, William D. 1979e. Communication in the physical and social sciences. In *Communication: The Essence of Science.* W. D. Garvey, pp. 280–299. New York: Pergamon Press.

Garvey, William D., and S. D. Gottfredson. 1979. Changing the system: Innovations in the interactive social system of scientific communication. In *Communication: The Essence of Science.* W. D. Garvey, pp. 300–321. New York: Pergamon Press.

Garvey, William D., and Belver C. Griffith. 1979a. Communication and information processing within scientific disciplines: Empirical findings for psychology. In *Communication: The Essence of Science.* W. D. Garvey, pp. 127–147. New York: Pergamon Press.

Garvey, William D., and Belver C. Griffith. 1979b. Scientific communication as a social system. In *Communication: The Essence of Science,* W. D. Garvey, ed., pp. 148–164. New York: Pergamon Press.

Garvey, William D., and Belver C. Griffith. 1963. *The American Psychological Association's Project on Scientific Information Exchange in Psychology.* Report no. 9. Washington, DC: American Psychological Association.

Garvey, William D., and Belver C. Griffith. 1972. Communication and information processing with scientific disciplines: Empirical findings for psychology. *Information Storage and Retrieval* 5: 123–136.

Garvey, William D., and Belver C. Griffith. 1971. Scientific communication: Its role in the conduct of research and creation of knowledge. *American Psychologist* 26(4): 14.

Garvey, William D., and Belver C. Griffith. 1968. Informal channels of communication in the behavioral sciences: Their relevance in the structure of formal or bibliographic communication. In *The Foundation of Access to Knowledge.* E. B. Montgomery, ed., pp. 129–146. Syracuse, NY: Syracuse University Press.

Garvey, William D., and Belver C. Griffith. 1967. Scientific communication as a social system. *Science* 157: 1011–1016.

Garvey, William D., Nan Lin, and Carnot E. Nelson. 1970. Communication in the physical and the social sciences. *Science* 170: 1166–1173.

Garvey, William, Nan Lin, and Kazuo Tomita. 1972. Research studies in patterns of scientific communication, III: Information-exchange processes associated with the production of journal articles. *Information Storage and Retrieval* 8: 207–221.

Garvey, William D., Kazuo Tomita, and Patricia Woolf. 1974. The dynamic scientific information user. In *Information Storage and Retrieval* 10: 115–131. Elmsford, NY: Pergamon Press.

Garvey, William, Nan Lin, Carnot E. Nelson, and Kazuo Tomita. 1978. *The Role of the National Meeting in Scientific and Technical Communication.* Baltimore: Johns Hopkins University Press.

Garvey, William, Nan Lin, Carnot E. Nelson, and Kazuo Tomita. 1972. Research studies in patterns of scientific communication, I: General description of research program. *Information Storage and Retrieval* 8(3): 111–122.

Garvey, William, Nan Lin, Carnot E. Nelson, and Kazuo Tomita. 1979. Research studies in patterns of scientific communication, I: General description of research program. In *Communication: The Essence of Science.* W. D. Garvey, ed., pp. 165–183. New York: Pergamon Press.

Garvey, William D., Kazuo Tomita, and Patricia Woolf. 1968. The dynamic scientific information user. *Information Storage and Retrieval* 15 (December 1968): 115–131.

Garvey, William D., et al. 1970a. Communciation in the physical and social sciences. *Science* 170(3963): 1166–1173.

Garvey, William D., et al. 1970b. Some comparisons of communication activities in the physical and social sciences. In *Communication Among Scientists and Engineers.* C. E. Nelson and D. K. Pollock, eds., pp. 61–64.

Gellman, Aaron J., and Stephen Feinman. 1975. The role and applications of scientific and technical information (STI) in the process of innovation and conception. Washington, DC: National Science Foundation. (Available NTIS PB256580.)

Gerstberger, Peter G., and Thomas J. Allen. 1968. Criteria used by research and development engineers in the selection of an information source. *Journal of Applied Psychology* 52(4): 272–279.

Gerstenfeld, Arthur, and Paul Berger. 1980. An analysis of utilization differences for scientific and technical information. *Management Science* 26(2): 165–179.

Gerstl, J. E., and J. P. Hutton. 1966. *Engineers: The Anatomy of a Profession.* London: Tavistock Publications.

Gessesse, Kebede. 1994. Scientific communication, electronic access, and document delivery: The new challenges to the science/engineering reference librarian. *International Information and Library Review* 26: 341–349.

Getz, R. R. 1978. A survey of New Jersey's agricultural weather service users. *Bulletin of the American Meteorological Society* 59(10): 1297–1304.

Gibson, R. 1973. Project management information requirements in an international organization. In *Project Management and Project Control, 10th ESRO*

Summer School, Frascati, Italy, September 1972, pp. 349–352. ESRO SP–90.

Gilchrist, Alan. 1983. Information provision for civil engineers: A pilot study. *British Library R&D Report* 5761 (April 1983): 83 pages.

Giuliano, V. E. 1981a. Teleworking: A prospectus. *Telephony* 200: 2 (January 12): 67, 70–72, 75.

Giuliano, V. E. 1981b. Teleworking: Future shock? *Telephony* 200: 6 (February 9): 56, 58, 60, 62.

Glaser, Edward M., and Samuel H. Taylor. 1973. Factors influencing the success of applied research. *American Psychologist* 23(2): 140–146.

Glass, B., and S. H. Norwood. 1959. How scientists actually learn of work important to them. In *Proceedings of the 1958 International Conference on Scientific Information,* pp. 195–197. Washington, DC: NAS.

Glassman, Myron, and Nanci A. Glassman. 1981. A review and evaluation of the Langley Research Center's scientific and technical information program: Results of Phase IV—Knowledge and Attitudes Survey, Academic and Industrial Personnel. Washington, DC: National Aeronautics and Space Administration. NASA TM-81894. (Available from NTIS 81N22976.)

Glassman, Nanci A., and Thomas E. Pinelli. 1992. An initial investigation into the production and use of scientific and technical information (STI) at five NASA centers: Results of a telephone survey, report 12. Washington, DC: National Aeronautics and Space Administration. NASA TM-104173. (Available from NTIS: 92N27170.)

Glock, C. Y., and H. Menzel. 1966. *The Flow of Information Among Scientists.* New York: Columbia University.

Glueck, William F., and Lawrence R. Jauch. 1975. Sources of research ideas among productive scholars. *Journal of Higher Education* 46(1): 103–114.

Goetzinger, C., and M. Valentine. 1964. Problems in executive interpersonal communication. *Personnel Administration* 27(2): 24–29.

Goldhar, J. D., et al. 1976. Information flows, management styles, and technological innovation. *IEEE Transactions on Engineering Management* EM-23 (1): 51–62.

Goodings, Deborah J., and Stephen A. Ketcham. 2001. Research versus practice in transportation geotechnics: Can we bridge the chasm? *Journal of Professional Issues in Engineering Education and Practice* 127(1): 26–31.

Goodman, Arnold F., et al. 1965. *DOD User Needs Study: Phase 2.* Anaheim, CA: North American Aviation. Available from NTIS AD647111, AD647112.

Goodman, Irene F., et al. 2002. *Women's Experiences in College Engineering.* Cambridge, MA: Goodman Research Group.

Gould, Constance C., and Karla Pearce. 1991. *Information Needs in the Sciences: An Assessment.* Mountain View, CA: Research Libraries Group.

Graham, Warren R., Clinton B. Wagner, William P. Gloege, and Albert Zavala. 1966. *Exploration of Oral/Informal Technical Communications Behavior, Final Report.* Washington, DC: Advanced Research Projects Agency, Defense Research and Engineering.

Gralewska-Vickery, A. 1976. Communication and information needs of earth science engineers. *Information Processing and Management* 12(4): 251–282.

Gralewska-Vickery, A., and H. Roscoe. 1975. *Earth Science Engineers Communication and Information Needs: Final Report and Appendix.* Available from Imperial College Research Report No. 32.

Grant, S. J. C. 1981. Cornerstone of a multifunction workstation: Electronic mail. *Telecommunications* 15(7): 34F–34H, 71.

Gray, T. G. F. 1981. Communication skills in engineering. *CME* 28(4): 23–27.

Grazia, A. de. 1968. Response to W. D. Garvey. In *The Foundations of Access to Knowledge.* E. B. Montgomery, ed., pp. 147–151. Syracuse University Press.

Griffin, Abbie, and John R. Hauser. 1992. Patterns of communication among marketing, engineering and manufacturing: A comparison between two new product teams. *Management Science* 38(3): 360–373.

Griffin, Stephen M. 1998. NSF/DARPA/NASA Digital Libraries Initiative: A program manager's perspective. *D-Lib Magazine* 4(7/8). Available from World Wide Web: <http://www.dlib.org/dlib/july98/07griffin.html>.

Griffith, Belver C., ed. 1980. *Key Papers in Information Science.* White Plains, NY: Knowledge Industry Publications.

Griffith, B., and A. J. Miller. 1970. Networks of informal communication among scientifically productive scientists. In *Communication Among Scientists and Engineers.* C. E. Nelson and D. K Pollock, eds., pp. 125–140. Lexington, MA: Heath Lexington Books.

Griffith, Belver C., and Nicholas Mullins. 1992. Coherent social groups in scientific change. *Science* 177.

Griffiths, José-Marie, and Donald W. King. 1991. A manual on the evaluation of information centers and services. Prepared for NATO, AGARD, April 1991. (Available from American Institute of Aeronautics and Astronautics, Technical Information Service, 555 West 57th Street, Suite 1200, New York, 10019).

Griffiths, José-Marie, and Donald W. King. 1993. *Special Libraries: Increasing the Information Edge.* Washington, DC: Special Libraries Association.

Griffiths, J.-M., B. C. Carroll, Donald W. King, M. E. Williams, and C. M. Sheetz. 1991. *Description of Scientific and Technical Information in the U.S.: Current Status and Trends,* 1. Available from the University of Tennessee, School of Information Sciences.

Grigg, Karen Stanley. 1998. Information-seeking behavior of science and engineering researchers at the Environmental Protection Agency Library in Research Triangle Park, North Carolina: Factors influencing choices to perform a self-search or to request a mediated search (Master's paper). University of North Carolina.

Grossman, S. D., and R Lindhe. 1980. People react to more than the message. *Human Systems Management* 1(3): 261–267.

Grove, Laurel, Don Zimmerman, and Marilee Long. 1996. Bringing communication science to technical communication: The importance of theory and research methods. In *Proceedings of the 1996 IEEE International Professional Communication Conference,* pp. 223–230. Piscataway, NJ: IEEE.

Grunig, J. E. 1980. Communication of scientific information to non-scientists. In

Progress In Communication Science, B. Dervin and M. J. Voigt, eds., 2: 167–214. Norwood, NJ: Ablex Publishing.

Guetzkow, H. 1968. Differentiation of roles in task-oriented groups. In *Group Dynamics: Research and Theory,* 3rd ed. D. Cartwright and A. Zander, eds., pp. 512–526. Tavistock Publishers.

Guindon, Mary Hardesty. 1994. Understanding the role of self-esteem in managing communication quality. *IEEE Transactions on Professional Communication* 37(1): 21–27.

Gullahorn, J. T. 1952. Distance and friendship as factors in the gross interaction matrix. *Sociometry* 15: 123–134.

Gunn, Craig James. 1995. Engineering graduate students as evaluators of communication skill. In *1995 ASEE Annual Conference Proceedings,* pp. 287–290. Washington, DC: ASEE.

Gunn, Craig James. 1998. Addressing the communication needs of a mechanical engineering department. In *1998 ASEE Annual Conference Proceedings.* Washington, DC: ASEE.

Gupta, B. M. 1981. Information, communication and technology transfer: A review of literature. *Annals of Library Science and Documentation* 28(1–4): 1–13.

Gupta, B. M., Suresh Kumar, H. K. Khanna, and T. K. Amla. 1999. Impact of professional and chronological age on the productivity of scientists in engineering science laboratories of CSIR. *Malaysian Journal of Library and Information Science* 4(1): 103–107.

Gupta, R. C. 1988. Skill development to assess information needs and seeking behavior. *Lucknow Librarian* 20(2): 52–58.

Gupta, R. C. 1993a. Communication among structural engineers. In *Advances in Library and Information Science, Vol. 4: Informal Communication,* pp. 21–30. India: Scientific Publishers.

Gupta, R. C. 1993b. Information and communication needs of structural engineering researchers. In *Advances in Library and Information Science, Vol. 4: Informal Communication,* pp. 119–127. India: Scientific Publishers.

Hagedorn, Katrina. 1996. *Comparison of the Usage of and Access to Electronic and Print Journals at the Engineering Library at the University of Michigan.* Ann Arbor, MI: School of Information, University Library Associates Project.

Hagstrom, W. D. 1965. *The Scientific Community.* Basic Books.

Hailey, Jeffrey C. 2000. Effective communication for EMC engineers. In *Proceedings of the 2000 IEEE International Symposium on Electromagnetic Compatibility,* pp. 265–268. Piscataway, NJ: IEEE.

Halbert, M. H., and R. L. Ackoff. 1959. An operations research study of the dissemination of scientific information. In *Proceedings of the International Scientific Conference on Information* 1: 97–130. Washington, DC: NAS.

Hall, Angela M. 1972a. Comparative use and value of INSPEC services. London, England: Institution of Electrical Engineers, INSPEC (INSPEC report R 72/9; OSTI report 5145).

Hall, Angela M. 1972b. *INSPEC: User Preference in Printed Indexes.* London: Institution of Electrical Engineers.

Hall, A. M. 1973. Methodology and results of some user studies on secondary information services. In *EURIM: European Conference on Research into the Management of Information Services and Libraries*, pp. 31–37. ASLIB.

Hall, Angela, P. Clague, and T. M. Aitchison. 1972. *The Effect of the Use of an SDI Service on the Information-Gathering Habits of Scientists and Technologists*. London: Institution of Electrical Engineers.

Hall, Homer J. 1975. Generalized method for the user evaluation of purchased information services. Linden, NJ: EXXON Research and Engineering Co. Available from ERIC ED 121139.

Hallmark, Julie. 1995. The effects of technology on the information seeking behavior of scientists. In *Proceedings of the 29th Meeting of the Geoscience Information Society*, pp. 51–56. Alexandria, VA: Geoscience Information Society.

Handy, C. B. 1979. *Understanding Organizations*. New York: Penguin.

Hanley, K., John Harrington, and John Blagden. 1998. *Aerospace Information Management: Final Report*. Cranfield: Cranfield University Press.

Harkins, C. 1981. Mathematical considerations in seeking fundamental limits to professional communication processes and effects. In *Proceedings of IEEE* 69(2): 167–170. Piscataway, NJ: IEEE.

Harmon, J. E., and D. R. Hamrin. 1993. Bibliography on communicating technical research information. *IEEE Transactions on Professional Communications* 36(1): 2–6.

Harney, Mick. 2000. Is technical writing an engineering discipline? *IEEE Transactions on Professional Communication* 43(2): 210–212.

Harrington, John, and John Blagden. 1999. The neglected asset: Information management in the UK aerospace industry. *Business Information Review* 16(3): 128–136.

Harris, D. 2001. Supporting human communication in network-based systems engineering. *Systems Engineering* 4(3): 213–221.

Harris, D., and Candy, L. 1999. Evaluation in the SEDRES project: Measuring the effectiveness of model data exchange between system engineering tools. In *Proceedings and on CD-ROM of the International Council on Systems Engineering 9th Annual Symposium*, INCOSE '99, Brighton, England.

Harris, William J., Jr. 1969. Creative dissemination of technical information. In *The Engineering Manager: Survival in the Seventies, 17th Joint Engineering Management Conference*. Engineering Institute of Canada, Montreal, pp. 61–70.

Havelock, Ronald G. 1972. *Bibliography on Knowledge Utilization and Dissemination*. Washington, DC: National Aeronautics and Space Administration.

Hawkins, H. S. 1980. Training of agricultural scientists in scientific communication. *International Forum of Information and Documentation* 5(3): 30–35.

Hawthorne, Sian. 1999. Providing engineering information for educators. *The New Review of Information Networking* 5: 147–152.

Hecht, Laura M., John M. Kennedy, Thomas E. Pinelli, and Rebecca O. Barclay. 1994. The technical communications practices of aerospace engineering and science students: Results of the Phase 4 cross-national surveys, report 28.

Washington, DC: National Aeronautics and Space Administration. NASA TM-109123. (NTIS pending.)

Hecht, Laura M., John M. Kennedy, Thomas E. Pinelli, and Rebecca O. Barclay. 1994. The technical communications practices of engineering and science students: Results of the Phase 3 survey conducted at the University of Illinois, report 27. Washington, DC: National Aeronautics and Space Administration. NASA TM-109122. (NTIS pending.)

Hecht, Laura M., John M. Kennedy, Thomas E. Pinelli, and Rebecca O. Barclay. 1994. The technical communications practices of aerospace engineering students: Results of the Phase 3 AIAA National Student Survey, report 26. Washington, DC: National Aeronautics and Space Administration. NASA TM-109121, June 1994. (NTIS pending.)

Hensley, Susan E., and Carnot Nelson. 1979. *Information about Users and Uses: A Literature Review.* ED: 176–788.

Herbert, E. 1970. Innovation in communication: Shortening the feedback loops among humans. In *Innovations In Communication Conference,* pp. 170–183. CFSTI PB 19229412.

Herkert, Joseph R., and Christine S. Nielsen. 1998. Assessing the impact of shift to electronic communication and information dissemination by a professional organization: An analysis of the Institute of Electrical and Electronics Engineers (IEEE). *Technological Forecasting and Social Change* 57: 75–103.

Herkert, Joseph R., and Christine S. Nielsen. 1996. Identifying obstacles in the shift to electronic media by professional societies: A Delphi study of the IEEE. In *Proceedings of the 1996 International Symposium on Technology and Society.* New York: IEEE.

Herner, Saul. 1954. Information gathering habits of workers in pure and applied science. *Industrial and Engineering Chemistry* 46(1): 228–236.

Herner, Saul, and Mary Herner. 1967. Information needs and uses. In *Annual Review of Information Science and Technology* 2: 1–34. New York: Interscience.

Herner, Saul., et al. 1979. *An Evaluation of the Goddard Space Flight Center Library.* Herner and Co. NASA CR–159969.

Hernon, Peter, and Thomas E. Pinelli. 1992. Scientific and technical information (STI) policy and the competitive position of the U.S. aerospace industry, paper, 18. Paper presented at the *30th Aerospace Meeting of the American Institute of Aeronautics and Astronautics (AIAA),* January 1992, Reno, NV. (Available from AIAA 92A28233.)

Heroux, Ronald G. 1981. Issues of computer conferencing. *Proceedings of the Technological Transfer Society International Symposium,* June 1981, 6.3-1–6.3-3.

Hertzum, Morten. 1999. Six roles of documents in professionals' work. In *Proceedings of the Sixth European Conference on Computer Supported Cooperative Work,* pp. 41–60. Dordrecht: Kluwer.

Hertzum, Morten, and Annelise Mark Pejtersen. 2000. Information seeking practices of engineers: Searching for documents as well as for people. *Information Processing and Management* 36: 761–778.

Herzog, Eric, and Anders Törne. 1999. Towards a standardized systems engineering information model. *Proceedings 9th International Symposium of the International Council on Systems Engineering, 1, INCOSE 99.*

Herzog, A. J. 1983. Career patterns of scientists in peripheral communities. Research Policy 12: 6 (December): 341–349.

Herzog, A. J. 1976. The "gatekeeper" hypothesis and the international transfer of scientific knowledge. In *The Problem of Optimization of User Benefit in Scientific and Technological Information Transfer,* pp. 15.1–15.9. AGARD CP-179.

Hewins, Elizabeth T. 1990. Information need and use studies. In *Annual Review of Information Science and Technology* 25. Martha E. Williams, ed. Amsterdam: Elsevier Science Publishers.

Hiles, K. E., and V. Wilczynski. 1995. Writing and engineering: A natural and necessary association. In *1995 ASEE Annual Conference Proceedings,* pp. 409–412. Washington, DC: ASEE.

Hilgard, E. R., and G. H. Bower. 1975. *Theories of Learning,* 4th ed. Englewood Cliffs: Prentice-Hall.

Hill, R. John. 1989. Using personal bibliographic data bases to keep up with the engineering literature. *IEEE Transactions on Professional Communication* (IPC) 32(3): 189–193.

Hill, R., and B. J. White. 1980. *Matrix Organization and Project Management.* Ann Arbor: University of Michigan Press.

Hills, Phillip, ed. 1980. *The Future of the Printed Word.* London: Frances Pinter Publishers.

Hills, P. J. 1979. Information package on teaching and learning methods for librarians. British Library, BLR&D Report 5512.

Hinrichs, J. R. 1964. Communications activity of industrial personnel. *Personnel Psychology* 17: 193–204.

Hirsch, Herb L. 1994. Technical communications. *IEEE Potentials* 13(2): 4–7.

Hirsch, Penny L., et al. 2001. Engineering design and communication: The case for interdisciplinary collaboration. *International Journal of Engineering Education* 17(4/5): 342–348.

Hirsh, Sandra G. 1999. Information seeking at different stages of the R&D process. In *Proceedings of SIGIR '99,* pp. 285–286. New York: ACM.

Hoch, Paul K. 1987. Institutional versus intellectual migrations in the nucleation of new scientific specialties. *Studies in History and Philosophy of Science* 18(4): 481–500.

Hogg, I. H., and J. R. Smith. 1959. Information and literature use in a research and development organization. In *Proceedings of the 1958 International Conference on Scientific Information,* pp. 131–162. Washington, DC: NAS.

Holland, Maurita P., Thomas E. Pinelli, Rebecca O. Barclay, and John M. Kennedy. 1991. Engineers as information processors: A survey of U.S. aerospace engineering faculty and students, paper 20. Reprinted from *European Journal of Engineering Education* 16(4): 317–336. (Available from NTIS 92N28155.)

Holland, Maurita Peterson. 1998. Modeling the engineering information professional. *Science & Technology Libraries* 17(2): 31–43.

Holland, Maurita Peterson, and Christina Kelleher Powell. 1995. A longitudinal survey of the information seeking and use habits of some engineers. *College & Research Libraries* 56(1): 7–15.

Holland, Maurita Peterson, and Christina Kelleher Powell. 1996. Two goals, one course: Using library school students as research mentors. *Research Strategies* 14(4): 196–204.

Holland, W. E., et al. 1976. Information channel/source selection as a correlate of technical uncertainty in a research and development organization. *IEEE Transactions on Engineering Management* EM-23(4): 163–167.

Holland, W. E. 1972. Characteristics of individuals with high information potential in government research and development organizations. *IEEE Transactions on Engineering Management* EM-19(2): 38–44.

Hollis, Richard. 1998. Creating online communities on the Internet: A practical solution for the global engineering community. *IATUL Proceedings* 7. Available from World Wide Web: <http://educate.lib.chalmers.se/IATUL/proceed-contents/fullpaper/hollis.html>.

Holmes, P. L. 1977. *On-line Information Retrieval, Vol. 1. Experimental Use of Non-medical Information Services.* London: British Library.

Holmfeld, John D. 1970. Communication behavior of scientists and engineers Ph.D.: Case Western University.

Holst, P., et al. 1996. 1995 IEEE electronic communication questionnaire. *The Institute.* Available from World Wide Web: <http://www.institute.ieee.org/INST/jan96/ecsurvey. html>.

Holsti, O. R. 1968. Context analysis. In *The Handbook of Social Psychology, 2nd ed., Vol. 2: Research Methods.* G. Lindzey and E. Aronson, eds., pp. 596–692. Boston: Addison-Wesley.

Horváth, László, and Imre J. Rudas. 1999. Human intent description as a tool for communication between engineers. In *Proceedings of the 1999 IEEE International Conference on Systems, Man, and Cybernetics,* II-348—II-353. Piscataway, NJ: IEEE.

Hoschette, John A. 2000. The effect of the Internet on engineering careers. *iSensors* 1(1): 50–52. Available from World Wide Web: <http://www.sensorsmag.com/isensors/dec00/ 50/index.htm>.

Hoslett, D. S. 1969. Barriers to communication. In *Readings in Management,* 3rd ed. M. D. Richards and W. A. Nielander, eds., pp. 147–154. Cincinnati: South-Western Publishing Company.

Hosono, Kimio, and Makiko Miwa. 1982. Some aspects of the use of online database services in Japan. In *Proceedings of the 6th International Online Information Meeting.* Oxford, England: Learned Innovation, pp. 243–250.

Houghton, Bernard. 1975. *Scientific Periodicals.* Hamden, CT: Linnet Books.

Howton, F. W. 1962–3. Work assignment and interpersonal relations in a research organization: Some participant observations. *Administrative Science Quarterly* 7: 502–520.

Hoyt, J. W. 1962. Periodical readership of scientists and engineers in research and development laboratories. *IRE Transactions on Engineering Management* EM-9, no. 2 (June 1962): 71–75.

Huang, G. Q., W. Y. Yee, and K. L. Mak. 2001. Development of a web-based system for engineering change management. *Robotics and Computer-Integrated Manufacturing* 17(3): 255–267. London: Elsevier Science Ltd.

Huber, K. C. 1980. An overview of six widely distributed statistical packages. In *Multivariate Techniques in Human Communication Research.* P. R. Monge and J. N. Cappella, eds., pp. 529–540. New York: Academic Press.

Hull, C. H., and N. H. Nie. 1981. *SPSS Update 7–9.* New York: McGraw-Hill.

Hunter, J. F., and C. W. Shockley. 1992. NASA scientific and technical program: User survey. NASA Technical Memo 108979.

Hurd, Julie M., Ann C. Weller, and Karen L. Curtis. 1992. Information seeking behavior of faculty: Use of indexes and abstracts by scientists and engineers. *Proceedings of the American Society for Information Science* 29: 136–143.

Hurst, Alexandra, and Ron S. Blicq. 1994. Modular format teaching of engineering management communications. In *Proceedings of the 1994 IEEE International Professional Communications Conference,* pp. 101–105. Piscataway, NJ: IEEE.

Hurwitz, J. I., et al. 1968. Some effects of power on the relations among group members. In *Group Dynamics: Research and Theory,* 3rd ed. D. Cartwright and A. Zander, eds., pp. 291–297. London: Tavistock Publications.

Hutchinson, Robert A., Jack E. Eisenhauer, Gerald J. Hane, and Donna C. Debrodt. 1985. Information flow from Japan to U.S. researchers in applied and basic energy fields. *Journal of Technology Transfer* 10(1): 1–7.

IEEE. 1996. *Meeting Member Needs in the 21st Century: IEEE Strategies for the Future.* New York: IEEE, pp. 337–346.

Indik, B. P. 1961. Superior-subordinate relationships and performance. *Personnel Psychology* 14: 357–374.

Irwin, Harry, and Elizabeth More. 1991. Technology transfer and communication: Lessons from Silicon Valley, Route 128, Carolina's Research Triangle and Hi-Tech Texas. *Journal of Information Science* 17(5): 273–280.

Jackson, J. M. 1959. The organization and its communications problem. *Advanced Management* 24(2): 17–20.

Jacobson, E., and S. E. Seashore. 1951. Communication practices in complex organizations. *Journal of Social Issues* 7: 28–40.

Jahoda, Gerald, Alan Bayer, and William L. Needham. 1978. A comparison of on-line bibliographic searches in one academic and one industrial organization. *RQ* (Fall): 42–49.

Jahoda, M., et al. 1957. *Research Methods in Social Relations with Especial Reference to Prejudice, Part 2: Selected Techniques.* Dryden.

James, G. D. 1980. Analysis and interpretation of data. In *Data Handling for Science and Technology.* S. A. Rossmassler and D. G. Watson, eds., pp. 55–64. North-Holland.

Jankovic, G., and L. Black. 1996. Engineering a web site. *IEEE Spectrum* 33(11): 62–69.

Jarvenpaa, S. L., and D. S. Staples. 2000. The use of collaborative electronic media for information sharing: An exploratory study of determinants. *Journal of Strategic Information Systems* 9: 129–154.

Jenny, H. K. 1978. Heavy readers are heavy hitters. *IEEE Spectrum* 15(9): 66–68.

Jerkovsky, W. 1983. Functional assessment in matrix organizations. *IEEE Transactions on Engineering Management* EM-30(2): 89–97.

Johns Hopkins University. 1971. Center for Research in Scientific Communication. *The Information-Dissemination Process Associated with Journal Articles Published by Heating, Refrigerating and Air-Conditioning Engineers.* Baltimore, MD: Johns Hopkins University (JHU-CRSC 048 874).

Johnson, Alan W. 1989. Perceived barriers and opportunities to engineering technical communication in selected Air Force organizations. Master's thesis.

Johnson, J. F. E., F. Luise, J-L. Loeuillet, M. Inderst, B. Nilsson, A. Torne, L. Candy, and D. Harris. 1999. The future systems engineering data exchange standard STEP AP-233: Sharing the results of the SEDRES Project. *Proceedings 9th International Symposium of the International Council on Systems Engineering, 1, INCOSE 99,* pp. 923–931.

Johnson, P. 1996. *Changing the Culture: Engineering Education into the Future.* Canberra: Institution of Engineers.

Johnson, N. L., and F. C. Leone. 1964. *Statistics and Experimental Design on Engineering and the Physical Sciences,* 2 vols. New York: Wiley.

Johnston, Ron, and Michael Gibbons. 1975. Characteristics of information usage in technological innovation. *IEEE Transactions on Engineering Management* EM-22(1): 27–34.

Johnston, R. E., and G. V. Smith. 1980. Communication in today's engineering environment. *SAWE Journal* 39(3): 33–41.

Jones S. R., and P. J. Thomas. 1999. Nationality as a factor in the use of information management technologies. *Behaviour & Information Technology* 18(4): 231–233.

Kagan, J., and E. Havemann. 1980. *Psychology: An Introduction,* 4th ed. New York: Harcourt, Brace, and Jovanovich.

Kahin, Brian. 1994. A cooperative framework for enhancing research communication in science and technology. *Serials Review* 20(4): 17–19.

Kahn, R. L., and C. F. Connell. 1967. *The Dynamics of Interviewing: Theory, Techniques, and Cases.* New York: Wiley.

Kant, Raj, and Jon Krueger. 1992. *Engineering Information System (EIS).* Final report for period September 1987–July 1991. Wright-Patterson Air Force Base, OH: Manufacturing Technology Directorate, Wright Laboratory, Air Force Systems Command, 1992. (Available from NTIS AD-A254013.)

Kantor, Paul B. 1982. Evaluation of and feedback via information storage and retrieval systems. In *Annual Review of Information Science and Technology* 17. White Plains, NY: Knowledge Industry Publications.

Karon, Paul. 1992. What engineering managers can do to muster motivation: Communication is key in boosting productivity. *EDN* 37(14a): 1–2.

Kasperson, Conrad J. 1978. Psychology of the scientist: XXXVII. Scientific creativity: A relationship with information channels. *Psychological Reports* 42(3): 691–694.

Kasperson, Conrad J. 1978. An analysis of the relationship between information sources and creativity in scientists and engineers. *Human-Communication Research* 4(2): 113–119.

Kasperson, C. J. 1976. An exploratory analysis of information use by innovative, productive, and non-productive scientists and engineers. Ph.D. diss., Rensselaer Polytechnic Institute.

Katz, D., and R. L. Kahn. 1978. *The Social Psychology of Organizations,* 2nd ed. New York: Wiley.

Katz, Ralph. 1984. As research teams grow older. *Research Management* XXVII(1): 23–28.

Katz, Ralph. ed. 1988. *Managing Professionals in Innovative Organizations.* Cambridge, MA: Ballinger Publishing Company.

Katz, Ralph, and Michael Tushman. 1979. Communication patterns, project performance, and task characteristics: An empirical evaluation and integration in an R&D setting. *Organizational Behavior and Human Performance* 23(2): 139–162.

Katz, Ralph, and Michael L. Tushman. 1981. An investigation into the managerial roles and career paths of gatekeepers and project supervisors in a major R&D facility. *R&D Management* 11(3): 103–110.

Katz, Ralph, and Michael L. Tushman. 1983. A longitudinal study of the effects of boundary spanning supervision on turnover and promotion in research and development. *Academy of Management Journal* 26(3): 437–456.

Katzen, May. 1980. The changing appearance of research journals in science and technology: An analysis and a case study. In *Development of Science Publishing in Europe,* A. J. Meadows, ed. Amsterdam: Elsevier Science Publishers.

Kaufman, Harold G. 1983. *Factors Related to Use of Technical Information in Engineering Problem Solving.* Brooklyn: Polytechnic Institute of New York.

Kaula, P. N. 1991. Trends in information handling systems: Gatekeeper technology. *International Forum on Information and Documentation* 16(4): 9–14.

Kaye, Steve. 1998. Effective communication skills for engineers. *IIE Solutions* 30(9): 44–46.

Keenan, S. 1978. *Survey of Packages for Training Users of Online Information Services, Final Report.* CEC.

Keller, Robert T., and Winford E. Holland. 1975. Boundary-spanning roles in a research and development organization: An empirical investigation. *Academy of Management Journal* 18(2): 388–393.

Kemper, John D. 1990. *Engineers and Their Profession.* Philadelphia: Saunders.

Kendall, M. G. 1962. *Rank Correlation Methods,* 3rd ed. Griffin & Company.

Kennedy, John M., Thomas E. Pinelli, Laura F. Hecht, and Rebecca O. Barclay. 1994. An analysis of the transfer of scientific and technical information (STI) in the U.S. aerospace industry, paper 42. Paper presented at the *Annual Meeting of the American Sociological Association (ASA),* Los Angeles, CA, August 1994. (Available from AIAA.)

Kennedy, John M., Thomas E. Pinelli, and Rebecca O. Barclay. 1994. Technical communications in aerospace education: A study of AIAA student members, paper 40. Paper presented at the *32nd Aerospace Sciences Meeting of the American Institute of Aeronautics and Astronautics (AIAA),* Reno, NV, January 1994. (Available from AIAA 94-0858.)

Kennedy, John M., and Thomas E. Pinelli. The impact of a sponsor letter on mail survey response rates, paper 3. Paper presented at the *Annual Meeting of the American Association for Public Opinion Research,* May 1990, Lancaster, PA. (Available from NTIS 92N28112.)

Kennedy, S. 1981. *Marketing of Professional Information Services.* MCB Publications.

Kent, Allen, ed. 1989. *Encyclopedia of Library and Information Science* 44. New York: Marcel Dekker.

Kiesler, Sara B. 1987. Social aspects of computer environments. *Social Science* 72(1): 23–28.

Kim, Seung-Lye. 1998. Measuring the impact of information on work performance of collaborative engineering teams. In *Proceedings of the 1998 ASIS Midyear Meeting.* Available from World Wide Web: <http://www.asis.org/Conferences/MY98/Kim.htm>.

King, Donald W. 1968. Design and evaluation of information systems. In *Annual Review of Information Science and Technology* 3. Carlos Cuadra, ed. Chicago: Brittannica.

King, Donald W. 1980. Electronic alternatives to paper-based publishing in science and technology. In *The Future of the Printed Word.* Philip Hills, ed. London: Frances Pinter Publishers.

King, Donald W. 1977. Systemic and economic interdependencies in journal publication. *IEEE Transactions on Professional Communication* PC–20:2 (September 1977).

King, Donald W., and Edward C. Bryant. 1971. *Evaluation of Information Services and Products.* Washington, DC: Information Resources Press.

King, Donald W., with Jane Casto, and Heather Jones. 1994. *Communication by Engineers: A Literature Review of Engineers' Information Needs, Seeking Processes, and Use.* Washington, DC: Council on Library Resources.

King, Donald W., José -Marie Griffiths, Nancy K. Roderer, Ellen A. Sweet, and Robert R. V. Wiederkehr. 1982. *The Value of the Energy Data Base.* King Research Inc. Available from NTIS DE82-014250.

King, Donald W., F. W. Lancaster, D. D. McDonald, N. K. Roderer, and B. L. Wood. 1976. *Statistical Indicators of Scientific and Technical Communication,* Vol. I: A Summary Report. Available from GPO 083-000-00295-3.

King, Donald W., D. M. Liston, G. L. Kutner, and R. G. Havelock. 1985. *Analysis of Technology Assistance Available to Small High Technology Firms.* Prepared for Small Business Administration.

King, Donald W., Dennis D. McDonald, and Candace H. Olsen. 1978. *A Survey of Readers, Subscribers, and Authors of the Journal of the National Cancer Institute.* National Cancer Institute.

King, Donald W., Dennis D. McDonald, and Nancy K. Roderer. 1981. *Scientific Journals in the United States: Their Production, Use, and Economics.* New York: Academic Press.

King, Donald W., D. D. McDonald, N. K. Roderer, C. G. Schell, C. G. Schuller, and B. L. Wood. 1977. *Statistical Indicators of Scientific and Technical Communication* 1977 ed. Available from NTIS PB-278279.

King, Donald W., and Carol Hansen Montgomery. 2002. After migration to an electronic journal collection. *D_Lib Magazine* 8(12). Available from World Wide Web <*http://www.dlib.org/dlib/december02king/12king.html*>.

King, Donald W., Peggy W. Neel, and Barbara L. Wood. 1972. *Comparative Evaluation of the Retrieval Effectiveness of Descriptor and Free-text Search Systems Using CIRCOL.* Rockville, MD: Westat Research. Available from NTIS: AD 738 299.

King, Donald W., and Vernon E. Palmour. 1974. User behavior. In *Changing Patterns in Information Retrieval.* Carol Fenichel, ed. Philadelphia: ASIS.

King, Donald W., and Nancy K. Roderer. 1978. *Systems Analysis of Scientific and Technical Communication in the U.S.: The Electronic Alternative to Communication through Paper-based Journals.* Report to the National Science Foundation. King Research. Available from NTIS PB-281847.

King, Donald W., and Nancy K. Roderer. 1982. Communication in Physics: The Use of Journals, *Physics Today* 35(10): 43, 45.

King, Donald W., and Carol Tenopir. 1998. Economic cost models of scientific scholarly journals. In *Economic, Read Costs and Benefits of Electronic Publishing in Science—A Technical Study: Proceedings of the ICSU Press Workshop.* Available from World Wide Web: <http://www.bodley.ox.ac.uk/icsu/kingppr.htm>.

King, Donald W., and Carol Tenopir. 2001. Using and reading scholarly literature. *Annual Review of Information Science and Technology* 34. Martha Williams, ed., Medford, NJ: Information Technology Today.

King, Donald W., and Carol Tenopir. Scholarly journal and digital database pricing: Threat or opportunity. Chapte 3 in Jeffrey Mackie-Mason, ed. Cambridge, MA: MIT Press, 2004, in progress.

King, Donald W., Carol Tenopir, Carol Hansen Montgomery, and Sarah E. Aerni. 2003. Patterns of use by faculty at three diverse universities. *D-Lib Magazine* 9(10). Available from World Wide Web: <http://www.dlib.org/dlib/october03/king/10king.html>.

King Research, Inc. 1977. *Library Photocopying in the United States: With Implications for the Development of a Copyright Royalty Payment Mechanism.* Washington, DC: U.S. Government Printing Office.

King, William R., and Gerald Zaltman, eds. 1979. *Marketing Scientific and Technical Information.* Boulder, CO: Westview Press.

Klapsis, M. P., and V. Thomson. 1997. Information transfers as a metric for engineering processes. *IFIP TC5 International Conference on Computer Applications in Production and Engineering (CAPE'97),* pp. 248–58. London: Chapman & Hall.

Kleinman, Larry. 1983. The engineer and his future with the computer. *Specifying Engineer* 50: 37–39, 202–203.

Kling, Rob, and Geoffrey McKim. 2000. Not just a matter of time: Field differences and the shaping of electronic media in supporting scientific communication. *Journal of the American Society for Information Science* 51(14): 1306–1320.

Kochen, F., and H. Tagaliscozzo. 1974. Matching authors and readers of scientific papers. *Information Storage and Retrieval* 10: 197–210.

Koehler, J. W., K. W. E. Anatol, and R. L. Applebaum. 1981. *Organizational Communication: Behavioral Perspectives,* 2nd ed. New York: Holt, Rinehart and Winston.

Koehn, Enno. 1995. Interactive communication in civil engineering classrooms. *Journal of Professional Issues in Engineering Education and Practice* 121(4): 260–261.

Koehn, Enno. 2001. Assessment of communications and collaborative learning in civil engineering education. *Journal of Professional Issues in Engineering Education and Practice* 127(4): 160–165.

Kohl, John R., Rebecca O. Barclay, Thomas E. Pinelli, Michael L. Keene, and John M. Kennedy. 1993. The impact of language and culture on technical communication in Japan, paper 25. Reprinted from *Technical Communication* 40(1) (First Quarter, February 1993): 62–73. Available from NTIS 93N18592.

Kolodny, H. F. 1980. Matrix organization designs and new product success. *Research Management* 23(5): 29–33.

Korfhage, Robert R. 1974. Informal communication of scientific information. *Journal of the American Society for Information Science* (January–February, 1974): 25–32.

Kornfield, W. A., and C. E. Hewitt. 1981. The scientific community metaphor. *IEEE Transactions on Systems, Man and Cybernetics* SMC-11(1): 24–33.

Krall, George F., and Sandra L. Burgoon. 1976. Electronic storage and delivery of handbook-type information: An emerging new tool for engineers. *Current Research on Scientific and Technical Information Transfer,* Micropapers edition. New York: Jeffrey Norton Publishers.

Kranzberg, Melvin. 1976. Formal versus informal communication among researchers. In *Current Research on Scientific and Technical Information Transfer.* Micropapers edition, New York: Jeffrey Norton Publishers.

Kraus, D. C., and A. K. Gramopadhye. 2001. Effect of team training of aircraft maintenance technicians: Computer-based training versus instructor-based training. *International Journal of Industrial Ergonomics* 27(3): 141–157.

Krause, D. C. 1978. *Marine Science Networks Involving the Individual and UNESCO; Vol 1: The Intergovernmental Framework; Vol 2: Cooperation and Coordination.* UNESCO.

Krawitz, L., and H. Newhouse. 1978. A survey of external users of NWS information. *Bulletin of the American Meteorological Society* 59(10): 1288–1296.

Kreitz, P. A., L. Addis, H. Galic, and T. Johnson. 1996. *The Virtual Library in Action: Collaborative International Control of High-Energy Physics Preprints.* Stanford, CA: Stanford Linear Accelerator Center, Stanford University.

Kremer, Jeanette Marguerite. 1980. *Information Flow among Engineers in a Design Company* Ph.D. diss., University of Illinois, Urbana-Champaign.

Kreth, Melinda L. 2000. A survey of the co-op writing experiences of recent engineering graduates. *IEEE Transactions on Professional Communication* 43(2): 137–152.

Krikelas, James. 1993. Information-seeking behavior: Patterns and concepts. *Drexel Library Quarterly* 19(2): 5–20.

Krockel, H. 1990. Advanced materials data systems for engineering. In *Scientific and Technical Data in a New Era: Proceedings of the 11th International CO-DATA Conference*, p. 62. New York: Hemisphere.

Kuhn, Allan D., and Gladys A. Cotter. 1986. The DOD Gateway Information System (DGIS): User interface design. In *Proceedings of the 49th ASIS Annual Meeting* 23: 150–157.

Kuhn, M. R., and K. Vaught-Alexander. 1994. Context for writing in engineering curriculum. *Journal of Professional Issues in Engineering Education and Practice* 120: 392–400.

Kuhlthau, Carol Collier. 1988. Longitudinal case studies of the information search process of users in libraries. *Library and Information Science Research* 10(3): 257–304.

Kulthau, Carol Collier. 1993. A principle of uncertainty for information seeking. *Journal of Documentation* 45: 339–355.

Lacy, William B., and Lawrence Busch. 1983. Informal scientific communication in the agricultural sciences. *Information Processing & Management* 19(4): 193–202.

Ladendorf, J. M. 1970. Information flow in science, technology and commerce: A review of the concepts of the sixties. *Special Libraries* 61(5): 215–222.

Lalitha, M. 1995. Information seeking behavior of medical and engineering personnel: A comparative study with reference to their library use. *Library Science with a Slant to Documentation and Information Studies* 32(2): 65–74.

Lamb, B. C. 1994. *A National Survey of Communication Skills of Young Entrants to Industry and Commerce.* London: The Queen's English Society.

Lancaster, F. W. 1974. Assessment of the technical information requirements of users. In *Contemporary Problems in Technical and Library Information Center Management*, pp. 59–85. Alan Rees, ed. Washington, DC: American Society for Information Science.

Lancaster, F. W. 1978. *Towards Paperless Information Systems.* New York: Academic Press.

Lancaster, F. W. 1993. *If You Want to Evaluate Your Library*, 2nd ed. Champaign, IL: University of Illinois.

Landau, Herbert B., Jerome T. Maddock, Floyd F. Shoemaker, and Joseph G. Costello. 1983. An information transfer model to define information users and outputs with specific applications to environmental technology. *Journal of the American Society for Information Science* 33(2): 82–96.

Langrod, G. 1963. *The International Civil Service: Its Origins, Its Nature, Its Evolution.* Sijthoff.

Large, J. A. 1980. *The Foreign-Language Barrier: Problems in Scientific Communication.* Andre Deutsch.

Lawal, Ibironke. 2002. Scholarly communication: The use and non-use of e-print archives for the dissemination of scientific information. *Issues in Science and*

Technology Librarianship. Fall 2002. Available from World Wide Web: <http://www.istl.org/02-fall/article3.html>

Lawrence, P. R., and J. W. Lorsch. 1967. *Organization and Environment: Managing Differentiation and Integration.* Harvard UP.

Lazinger, Susan S., Judit Bar-Ilan, and Bluma C. Peritz. 1997. Internet use by faculty members in various disciplines: A comparative case study. *Journal of the American Society for Information Science* 48(6): 508–518.

Le Vie, Donald S., Jr. 1996. Skills, knowledge, and training for the 21st century technical communicator. In *Proceedings of the 1996 IEEE International Professional Communication Conference,* pp. 196–202. Piscataway, NJ: IEEE.

Le Vie, Donald S., Jr. 1997. How to become a value-add technical communicator to scientists, engineers, and technical staff. In *Proceedings of the Society for Techincal Communicators 44th Annual Conference,* pp. 466–467. Toronto: Interact Toronto.

Leary, T. 1961. The theory and measurement methodology of interpersonal communication. In *The Planning of Change,* W. G. Bennis, et al., pp. 307–321. New York: Holt, Rinehart and Winston.

Leavitt, H. J. 1951. Some effects of certain communication patterns on group performance. *Journal of Abnormal and Social Psychology* 46: 38–50.

Leckie, Gloria J., Karen E. Pettigrew, and Christian Sylvain. 1996. Modeling the information seeking of professionals: A general model derived from research on engineers, health care professionals, and lawyers. *Library Quarterly* 66(2): 161–193.

Lee, A. M. 1983. *Electronic Message Transfer and Its Implications.* Lexington, MA: Lexington Books.

Lee, Denis M. S. 1994. Social ties, task-related communication and first job performance of young engineers. *Journal of Engineering and Technology Management* 11: 203–228.

Lee, Dong Ho, and D. Richard Decker. 1994. Collaborative engineering-design support system. In *Proceeding of the 3rd Clips Conference* 2: 285–295. Houston: Johnson Space Center.

Leggett, Robert G. 1976. Do engineers read (or buy?) books? *Scholarly Publishing* 7(4): 337–342.

Leibson, David E. 1981. How Corning designed a "talking" building to spur productivity. *Management Review* 70: 8–13.

Lemoine, W. 1991. Industrial research output and productivity according to age. *Research Evaluation* 1: 161–174.

Leonard-Barton, Dorothy. 1990. The interorganizational environment: Point-to-point versus diffusion. In *Technology Transfer: A Communication Perspective.* Frederick Williams and David V. Gibson, eds., pp. 43–62. London: Sage Publications.

Lescoheir, R. S., M. A. Lavin, and M. K. Landsberg. 1984. Database development and end-user searching: Exxon Research and Engineering Company. *Science and Technology Libraries* 5(1): 1–15.

Levin, R. I., and D. S. Rubin. 1980. *Applied Elementary Statistics.* Englewood Cliffs, NJ: Prentice-Hall.

Levinson, Nanette S., and David D. Moran. 1987. R&D management and organizational coupling. *IEEE Transactions on Engineering Management* EM-34, no. 1 (February 1987): 28–35.

Levitt, S. R., and R. S. Howe. 2000. Visual and statistical thinking: An essential communication curriculum for engineers. In *Proceedings of the 3rd Working Conference on Engineering Education for the 21st Century,* pp. 155–158. Sheffield, England: Sheffield Hallam University Press.

Lewis, Barbara. 1993. Uses of language in collaborative design terms. In *Proceedings of the 1993 IEEE International Professional Communication Conference,* pp. 5–10. Piscataway, NJ: IEEE.

Li, Harry W. 1994. Engineering communication styles in a nonengineering world. In *Proceedings of the 1994 Frontiers in Education Conference,* pp. 486–490. Piscataway, NJ: IEEE.

Liebscher, Peter, Eileen G. Abels, and Daniel W. Denman. 1997. Factors that influence the use of electronic networks by science and engineering faculty at small institutions, Part II. Preliminary use indicators. *Journal of the American Society for Information Science* 48(6): 496–507.

Lievrouw, L. A. 1990. Communication and the social representation of scientific knowledge. *Critical Studies in Mass Communication* 7: 1–10.

Lievrouw, Leigh A., and Kathleen Carley. 1990. Changing patterns of communication among scientists in an era of "telescience." *Technology in Society* 12(4): 457–477.

Likert, R. 1961. *New Patterns of Management.* New York: McGraw-Hill.

Likert, R. 1967. *The Human Organization: Its Management and Value.* New York: McGraw-Hill.

Lin, Nan. 1973. *The Study of Human Communication.* Bobbs-Merrill.

Lin, Nan, and William D. Garvey. 1971. Stratification of the formal communication system in science. In *American Sociological Association Annual Meeting, 1971, August/September,* Denver, CO, Abstract, 137. AMA.

Lin, Nan, and William D. Garvey. 1972. Information needs and uses. In *Annual Review of Information Science and Technology* 7: 5–37. Washington, DC: American Society for Information Science.

Lin, Nan, William D. Garvey, and Carnot E. Nelson. 1970. A study of the communication structure of science. In *Communication Among Scientists and Engineers.* Carnot E. Nelson and Donald K. Pollack, eds., pp. 23–60. Lexington, MA: D.C. Heath.

Lindzey, G., and E. Aronson. 1968. *The Handbook of Social Psychology, 2nd ed., Vol. 2, Research Methods.* Reading, MA: Addison-Wesley.

Lipetz, Ben-Ami. 1970. Information needs and uses. In *Annual Review of Information Science and Technology* 5: 1–32. Chicago: Encyclopaedia Britannica.

Little, S. B. 1989. The research and development process and its relationship to the evolution of scientific and technical literature: A model for teaching research. *The Technical Writing Teacher* 16: 68–76.

Litwin, G. H., and R. A. Stringer. 1968. *Motivation and Organizational Climate.* Boston: Harvard University Graduate School of Business Administration.

Lloyd, Peter. 2000. Storytelling and the development of discourse in the engineering design process. *Design Studies* 21(4): 357–373.

Lloyds, R. M., C. J. Moore, and N. Kitching. 2001. Online CPD for engineers. In *IEE International Symposium Engineering Education: Innovations in Teaching, Learning, and Assessment, Vol. 1,* January 4–5, 2001. London, UK: IEE.

Llull, Harry P. 1991. Meeting the academic and research information needs of scientists and engineers in the university environment. *Science and Technology Libraries* 11(3): 83–90.

Longo, Bernadette. 1997. Who makes engineering knowledge? Changing identities of technical writers in the 20th century United States. In *Proceedings of the 1997 IEEE International Professional Communication Conference,* pp. 61–68. Piscataway, NJ: IEEE.

Lorenz, Patricia. 1980. Searches conducted for engineers. Paper presented at the *National Online Information Meeting,* New York, March 25–27, 1980. Arlington, VA: ERIC Document Reproduction Service (Computer Microfilm International Corporation).

Lowry, Glenn R. 1979. *Information Use and Transfer Studies: An Appraisal.* Available from ERIC ED-211085.

Luce, R. D., and A. D. Perry. 1949. A method of matrix analysis of group structure. *Pscyhometrika* 14(1): 95–116.

Ludewig, C. W. 1965. Information needs of scientists in missile and rocket research as reflected by an analysis of reference question. MLS thesis, University of North Carolina.

Lufkin, J. M., and E. H. Miller. 1966. The reading habits of engineers: A preliminary survey. *IEEE Transactions on Education* E-9(4): 179–182.

Lund, C. P., and P. J. Jennings. 2001. Potential, practice and challenges of tertiary renewable energy education on the World Wide Web. *Renewable Energy* 22(1): 119–125.

Mabe, Michael A. 2001. Digital dilemmas: Electronic challenges for the scientific journal publisher. In *ASLIB Proceedings* 53(3): 85–92.

Machlup, Fritz. 1980. *Knowledge—Its Creation, Distribution, and Economic Significance, Vol. 1: Knowledge and Knowledge Production.* Princeton: Princeton University Press.

Machlup, Fritz, and Kenneth W. Leeson. 1978. *Information Through the Printed Word: The Dissemination of Scholarly, Scientific, and Intellectual Knowledge, Vol. 2: Journals.* New York: Praeger.

Macleod, A., D. R. McGregor, and G. H. Hutton. 1994. Accessing of information for engineering design. *Design Studies* 15(3): 260–269.

MacLeod, Roddy, and Linda Kerr. 1997. EEVL: Past, present, and future. *Electronic Library* 15(4): 279–286.

Maguire, Carmel, Edward J. Kazlauskas, and Anthony D. Weir. 1994. *Information Science for Innovative Organizations.* New York: Academic Press.

Mahan, John E., Anura Jayasumana, Derek Lile, and Mike Palmquist. 2000. Bringing an emphasis on technical writing to a freshman course in electrical engineering. *IEEE Transactions on Education* 43(1): 36–42.

Mahé, Annaïg, Christine Andrys, and Ghislaine Chartron. 2000. How French research scientists are making use of electronic journals: A case study conducted at Pierre et Marie Curie University and Denis Diderot University. *Journal of Information Science* 26(5): 291–302.

Mailloux, Elizabeth N. 1989. Engineering information systems. In *Annual Review of Information Science and Technology* (24): 239–266. New York: Elsevier Science Publishers.

Majid, Shaheen, and Ai Tee Tan. 2002. Usage of information resources by computer engineering students: A case study of Nanyang Technological University, Singapore. *Online Information Review* 26(5): 318–325.

Majid, Shaheen, Tamara S. Eisenschitz, and Mumtaz Ali Anwar. 1999. Library use pattern of Malaysian agricultural scientists. *Libri* 49: 225–235.

Majid, Shaheen, Mumtaz Ali Anwar, and Tamara S. Eisenschitz. 2000. Information needs and information seeking behavior of agricultural scientists in Malaysia. *Library & Information Science Research* 22(2): 145–163.

Making the case for a national engineering information system: David Penniman has found his challenge. 1996. *InfoManage* 3(12): 1–5.

Manuel-Dupont, S. 1996. Writing-across-the-curriculum in an engineering program. *Engineering Education* 85: 35–40.

March, J. G., and Simon, H. A. 1958. *Organizations.* New York: Wiley.

Mardikian, Jackie. 1995. Science faculty perspectives: Meeting research needs in an electronic environment. *New Jersey Libraries* 28: 23, 25, 27.

Markusova, Valentina A., R. S. Gilyarevskii, and A. I. Chernyi. 1994. Communication among Russian scientists and between them and world science. *International Forum on Information and Documentation* 19(3/4): 17–27.

Markusova, Valentina A., R. S. Gilyarevskii, A. I. Chernyi, and Belver C. Griffith. 1996. Information behavior of Russion scientists in the "Perestroika" period: Results of the questionnaire survey. *Scientometrics* 37(2): 361–380.

Marquis, Donald G., and Thomas J. Allen. 1966. Communication patterns in applied technology. *American Psychologist* 21(11): 1052-1060.

Marsh, Hugh C. 1998. The engineer as document designer: The new world order. In *Proceedings of the 1998 Society for Technical Communication Annual Conference,* pp. 53–56. Arlington, VA: STC.

Marshall, Doris B. 1975. A survey of the use of on-line computer-based scientific search services by academic libraries. *Journal of Chemical Information and Computer Sciences* 15(4): 247–249.

Martin, M. P., and W. Fuerst. 1984. Communications framework for systems design. *Journal of Systems Management* 35(3): 18–25.

Martino, Al. 1993. Stone and Webster engineers phone system to bridge continents. *Communication News* 30(4): 32(1).

Martyn, John. 1974. Information needs and uses. In *Annual Review of Information Science and Technology* 9: 3–23. Washington, DC: American Society for Information Science.

Martyn, John. 1987. Literature searching habits and attitudes of research scientists, paper 14. British Library Research.

Mason, Robert M. 1977. *Development of a Cost Benefit Methodology for STI Com-*

munication and Applications to Information Analysis Centers. Metrics. Available from NTIS: PB 278 566/AS.

Matarazzo, J. M. 1981. *Closing the Corporate Library: Case Studies on the Decision-Making Process.* Special Libraries Association.

Mathieu, Richard G. 1995. The Internet: Information resources for industrial engineers. *Industrial Engineering* 27(1): 49–52.

Maxwell, A. E. 1961. *Analysing Qualitative Data.* Methuen.

McAlpine, A., et al. 1972. *The Flow and Use of Scientific Information in University Research.* OSTI Report 5138. Manchester Business School R&D Unit.

McCain, Katherine W. 2000. Sharing digitized research-related information on the World Wide Web. *Journal of the American Society for Information Science* 51(14): 1321–1327.

McClure, Charles R., Peter Hernon, and Gary R. Purcell. 1986. *Linking the National Technical Information Service with Academic and Public Libraries.* Norwood, NJ: Ablex Publishing Corporation.

McClure, C. R. 1989. Increasing access to U.S. scientific and technical information: Policy implications. In *U.S. Scientific and Technical Information (STI) Policies: Views and Perspectives,* C. R. McClure and P. Hernon, eds., p. 4. Norwood, NJ: Ablex.

McClure, C. R., A. Bishop, and P. Doty. Electronic networks, the research process, and scholarly communication: An empirical study with policy recommendations for the national research and education network. (Mimeograph.)

McClure, Charles R. 1987. Improving access to and use of federal scientific and technical information (STI): Perspectives from recent research projects. In *ASIS '87 Proceedings of the ASIS Annual Meeting,* Ching-chih Chen, ed., pp. 163–169. Medford, NJ: Learned Information.

McCullough, Robert A., Thomas E. Pinelli, Douglas D. Pilley, and Frede F. Stohrer. 1982. A review and evaluation of the Langley Research Center's Scientific and Technical Information Program. Washington, DC: National Aeronautics and Space Administration. NASA TM-83269.

McDonald, Dennis D., and Colleen G. Bush. 1982. *Libraries, Publishers and Photocopying: Final Report of Surveys Conducted for the U.S. Copyright Office.* King Research Report. Washington, DC: U.S. Copyright Office.

McDonald, K. A. 1991. Despite benefits, electronic journals will not replace print, experts say. *Chronicle of Higher Education* (February 27, 1991): A6.

McGregor, Helen. 2000. Engineers at work: Developing communication skills for professional practice. In *Proceedings of the 2000 Society for Technical Communication Annual Conference,* pp. 22–26. Arlington, VA: STC.

McGregor, Helen, and C. McGregor. 1998. Documentation in the Australian engineering workplace. *Australian Journal of Engineering Education* 8(1): 11–22.

Meadow, Charles T., Barbara A. Cerny, Christine L. Borgman, and Donald O. Case. 1989. Online access to knowledge: System design. *Journal of the American Society for Information Science* 40(2): 86–98.

Meadows, A. J. 1974. *Communication in Science.* London: Butterworths.

Meadows, A. J., and P. Buckle. 1993. Changing communication activities in the British scientific community. *Journal of Documentation* 48(3): 276–290.

Meadows, A. J., ed. 1979. *The Scientific Journal*. London: ASLIB.

Meadows, A. J., ed. 1980. *New Technology and Developments in the Communication of Research During the 1980s*. Primary Communications Research Centre.

Meadows, A. J. 1974. How the scientist acquires and uses information. In *Communication in Science*. A. J. Meadows, ed., pp. 91–125. London: Butterworths.

Meadows, A. J., and J. G. O'Connor. 1969. *An Investigation of Information Sources and Information Retrieval in Astronomy and Space Science*. 2 vols. Report No. 5044. OSTI.

Meadows, Jack. 1997. Changing patterns of communication and electronic publishing. *IATUL Proceedings* 7. Available from World Wide Web: <http://educate.lib.chalmers.se/ IATUL/proceedcontents/fullpaper/meadpap.html>.

Mehta, Usha, and Virginia E. Young. 1995. Use of electronic information resources: A survey of science and engineering faculty. *Science & Technology Libraries* 15(3): 43–54.

Meltzer, M. F. 1981. *Information: The Ultimate Resource*. Amacom.

Mench, J. W. 2002. Electrical education for construction engineers. In *Proceedings IEEE Southeast Conference 2002*, pp. 147–151. Columbia, SC: IEEE.

Menzel, H. 1960. *Review of Studies in the Flow of Information Among Scientists*. Bureau of Applied Social Research, Columbia University.

Menzel, H. 1962. Planned and Unplanned Scientific Communication. In B. Barber and Walter Hirsch, eds. *Sociology of Science*. New York: Macmillan (Free Press).

Menzel, H. 1964. The information needs of current scientific research. *Library Quarterly* 34: 4–19.

Menzel, H. 1966a. *Formal and Informal Satisfaction of the Information Requirements of Chemists*. Interim Report. Bureau of Applied Social Research. PB 173261. CFSTI.

Menzel, H. 1966b. Information needs and uses in science and technology. In *Annual Review of Information Science and Technology* 1: 41–69. Carlos Cuadra, ed. New York: Interscience.

Menzel, H. 1966c. Scientific communication: Five themes from social science research. *American Psychologist* 21: 1001.

Menzel, H. 1968. Informal communication in science: Its advantages and its formal analogues. In *The Foundations of Access to Knowledge*. E. B. Montgomery, ed., pp. 153–163. Syracuse UP.

Menzel, H. 1970. *Formal and Informal Satisfaction of the Information Requirements of Chemists*. Final report, Bureau of Applied Social Research. Columbia University. PB 193556. CFSTI.

Menzel, Herbert, L. Lieberman, and J. Dulchin. 1960. *Review of Studies in the Flow of Information Among Scientists*. New York: Columbia University Press.

Metayer-Duran, Cheryl. 1993. Information gatekeepers. In *Annual Review of Information Science and Technology* 28. Medford, NJ: Learned Information.

Michel, J. 1982. Linguistic and political barriers in the international transfer of information in science and technology. *Journal of Information Science* 5(4): 131–135.

Mick, C. K., et al. 1980. Towards usable user studies. *Journal of the Society for Information Science* 31(5): 347–356.

Mick, Colin K., Georg N. Lindsey, Daniel Callahan, and Frederick Spielberg. 1979. Towards usable user studies: Assessing the information behavior of scientists and engineers, December 1979. Washington, DC: National Science Foundation, Division of Information Science and Technology (NSF/IST78-10531F). Available NTIS PB80-177165.

Mikhailov, A. I., A. I. Chernyi, and R. S. Giliarevskii. 1984. *Scientific Communications and Informatics,* trans. R. H. Burger. Information Resources Press.

Miller, G. A. 1956. The magical number seven, plus or minus two: Some limits on our capacity for processing information. *Psychological Review* 63(2): 81–97.

Miller, Jeannie P., and Richard Stringer-Hye. 1995. Improved access to engineering society technical papers. *Reference Services Review* 23(3): 63–67.

Miller, R. L., and B. M. Olds. 1994. A model curriculum for a capstone course in multidisciplinary engineering design. *Journal of Engineering Education* 83: 311–316.

Mischo, William H., and Jounghyoun Lee. 1987. End-user searching of bibliographic databases. In *Annual Review of Information Science and Technology* 22. Martha E. Williams, ed. Medford, NJ: Learned Information.

Mitchell, T. R. 1982. *People in Organizations: An Introduction to Organizational Behavior.* 2nd ed. New York: McGraw-Hill.

Moenaert, Rudy K., Dirk Deschoolmeester, Arnoud De Meyer, and William E. Souder. 1992. Information styles of marketing and R&D personnel during technological product innovation projects. *R&D Management* 22(1): 21–38.

Mondschein, Lawrence G. 1990. Selective dissemination of information (SDI): Relationship to productivity in the corporate environment. *Journal of Documentation* 46(2): 137–145.

Mondschein, Lawrence G. 1990. Selective dissemination of information (SDI): Use and productivity in the corporate research environment. *Special Libraries Association* 81(4): 265–279.

Monge, Peter R., et al. 1979. The assessment of NASA technical information. communications. NASA CR-181367. 224 pages. Available NTIS 87N70843.

Monge, P. R., and J. N. Cappella. 1980. *Multivariate Techniques in Human Communications Research.* New York: Academic Press.

Montgomery, E. B. 1968. Four "new" sciences: An approach to complexity. In *AGARD Storage and Retrieval of Information: A User-Supplier Dialogue Conference,* June 18–20, 1968, Munich, West Germany, pp. 41–48. AD 481245 AGARD-CP-39. AGARD.

Montgomery, C. H., and D. W. King. 2002. Comparing library and user related costs of print and electronic collections. A first step towards a comprehensive analysis. *D-Lib Magazine* 8(10). Available from World Wide Web <*http://www. dlib.org/dlib/october02/montgomery/10montgomery.html*>.

Mooney, C. J. 1991. In 2 years, a million refereed articles, 300,000 books, chapters, monographs. *The Chronicle of Higher Education* (May 22, 1991): A17.

Moore, Ivan. 1996. Why do we have practicals anyway?: A focus on the essential engineering skills. In *Proceedings of the 1996 IEE Colloquium on Engineering Education in the Twenty-First Century,* 4/1–4/3. London: IEE.

Moravcsik, M. J. 1983. The role of science in technology transfer. *Research Policy* 12(5): 287–296.

Moroney, M. J. 1979. *Facts From Figures.* Penguin.

Morris, Ruth C. T. 1994. Toward a user-centered information center. *Journal of the American Society for Information Science* 45(1): 20–30.

Moser, C. A., and G. Kalton. 1977. *Survey Methods in Social Investigation,* 2nd ed. Heinemann.

Mosley, Pixie Anne. 1995. Engineers and librarians: How do they interact? *Science & Technology Libraries* 15(1): 51–61.

Mosteller, F., and J. W. Tukey. 1968. Data analysis including statistics. In *The Handbook of Social Psychology, 2nd ed., Vol. 2, Research Methods.* G. Lindzey and E. Aronson, eds., pp. 80–203. Reading, MA: Addison-Wesley.

Mucchielli, R. 1979. *Le Questionnaire Dans L'enquete Psycho-Sociale,* 6th ed. Entreprise Moderne d'Editions.

Muchinsky, P. M. 1976. An assessment of the Litwin and Stringer organization climate questionnaire: An empirical and theoretical extension of the Sims and Lafollette study. *Personnel Psychology* 29: 371–392.

Murphy, Daniel J. 1984. Computer mediated communication (CMC) and the communication of technical information in aerospace, paper 38. Paper presented at the *32nd Aerospace Sciences Meeting of the American Institute of Aeronautics and Astronautics (AIAA),* Reno, NV, January 1984. Available from AIAA 94-0840.

Murphy, M. 1979. Measuring library effectiveness: A prelude to change. *Special Libraries* 70(1): 18–25.

Myers, L. A. 1983. Information systems in research and development: The technological gatekeeper reconsidered. *R&D Management* 13(4): 199–206.

Myers, M. T., and G. E. Myers. 1982. *Managing by Communication: An Organizational Approach.* New York: McGraw-Hill.

Mykytyn, Peter P., Jr., Kathleen Mykytyn, and M. K. Raja. 1994. Knowledge acquisition skills and traits: A self-assessment of knowledge engineers. *Information & Management* 26: 95–104.

Nagle, Joan G. 1996. *Preparing Engineering Documents.* Piscataway, NJ: IEEE.

Nagle, Joan G. 1998. Communication in the profession. *Today's Engineer* 1(1).

Nagurney, Ladimer S., M. Saleh Keshawarz, and Ronald S. Adrezin. 2000. A junior course in engineering design and society. In *Proceedings of the 2000 Frontiers in Education Conference,* F4C-10. Piscataway, NJ: IEEE.

Nash, S. H. 1983. Selecting and implementing a word processor in the library. *The Electronic Library* 1(4): 265–273.

National Academy of Engineering. 1974. *Experiments in Research and Develop-*

ment Incentives. Final report. Washington, DC: National Academy of Engineering.

National Academy of Sciences. Committee on Scientific and Technical Communication. 1969. Scientific and technical communication: A pressing national problem and recommendation for its solution. Washington, DC: National Academy of Sciences.

Neale, Michael. 1982. A study of the potential contribution of the British Library to the information needs of engineers. Report TRB 293 (October 1982).

Nelson, Carnot E., and Donald K. Pollock. 1970. *Communication among Scientists and Engineers.* Lexington, MA: Heath Lexington Books.

Nelson, Michael L., Gretchen L. Gottlich, David J. Bianco, Sharon S. Paulson, Robert L. Binkley, Yvonne D. Kellogg, Chris J. Beaumont, Robert B. Schmunk, Michael J. Kurtz, Alberto Accomazzi, and Omar Syed. 1995. The NASA technical report server. *Internet Research: Electronic Networking Applications and Policy* 5(2): 25–36.

Nelson, R. C. 1968. Transcultural communication. In *Managerial Control Through Communication.* G. T. Vardaman and C. C. Halterman, eds., pp. 172–183. New York: Wiley.

Newcomb, T. M. 1953. An approach to the study of communicative acts. *Psychological Review* 60(6): 393–404.

Newell, Sue, and Peter Clark. 1990. The importance of extra-organizational networks in the diffusion and appropriation of new technologies: The role of professional associations in the United States and Britain. *Knowledge: Creation, Diffusion, Utilization* 12(2): 199–212.

Nicholas, D. 1996. An assessment of the online searching behavior of practitioner and end users. *Journal of Documentation* 52(3): 227–251.

Nie, N. H., et al. 1970. *SPSS: Statistical Package for the Social Sciences.* New York: McGraw-Hill.

Nie, N. H. 1975. *SPSS: Statistical Package for the Social Sciences,* 2nd ed. New York: McGraw-Hill.

Nochur, K. S., and Thomas J. Allen. 1992. Do nominated boundary spanners become effective technological gatekeepers? *IEEE Transactions on Engineering Management* 39(3): 265–269.

Nojima, Sheryl E. 1998. Revitalization of the engineering curriculum: Enhancing learning through improving communication and teamwork (ELICIT). In *Proceedings of the American Power Conference 1998 Annual Meeting,* pp. 578–579. Chicago: Illinois Institute of Technology.

Nosok, B. D., and A. A. Gorlov. 1974. Ob izuchenii potrebnostej vrachei tsentral'nykh bol'nits v nauchnoi meditsinskoi informatsii (Scientific medical information needs of district hospital staff: A user study). *Nauchno. Tekh. Inf.* 1(2): 11.

O'Flaherty, John J. 1997. EURILIA: European Initiative in Library and Information in Aerospace. *Interlending & Document Supply* 25(4): 157–165.

Oen, Carol, and Helen A. Pfuderer. 1982. Scientific and technical information for research and development. In *Proceedings of the 11th ASIS Midyear Meeting,* Knoxville, TN, June 13–16, 1982.

Olson, E. E. 1977a. Organizational factors affecting information flow in industry. *ASLIB Proceedings* 29: 2–11.

Olson, E. E. 1977b. Organizational factors affecting the flow of scientific and technical information in industrial R&D divisions. *NTIS*. PB 277761.

Oppenheim, A. N. 1968. *Questionnaire Design and Attitude Measurement.* Heinemann.

Orr, J. M. 1977. *Libraries as Communication Systems.* Greenwood Press.

Orr, R. H. 1968. (Response to H. Menzel). In *The Foundations of Access to Knowledge.* E. B. Montgomery, ed., pp. 164–167. Syracuse UP.

Orr. R. H. 1970. The scientist as an information processor: A conceptual model illustrated with data on variables related to library utilization. In *Communication Among Scientists and Engineers.* C. E. Nelson and D. E. Pollock, eds., pp. 143–189. Lexington, MA: Heath Books.

Osorio, Nestor L. 2001. Web sites of science-engineering libraries: An analysis of content and design. *Issues in Science and Technology Librarianship* 29.

Othway, H. J., and M. Peltu. 1983. *New Office Technology: Human and Organizational Aspects.* Francis Pinter.

Owen, David. 1999. Chemical reaction. *Engineering E2* (November 26): 10.

Packer, K. H. 1975. *Methods used by chemists and chemical engineers in Candian universities to maintain current awareness with special reference to the use of SDI systems.* Ph.D. diss., University of Maryland, Baltimore.

Page, Gillian, Robert Campbell, and Jack Meadows. 1987. *Journal Publishing: Principles and Practices.* Boston: Butterworths.

Paisley, W. J. 1965. *The Flow of (Behavioral) Science Information: A Review of the Research Literature.* Stanford University, Institute for Communication Research.

Paisley, William J. 1968. Information needs and uses. In *Annual Review of Information Science and Technology* 3. Chicago: Brittannica.

Paisley, W. 1980. Information and work. In *Progress and Communication Sciences, Vol. 11.* B. Dervin and M. J. Voigt, eds., pp. 113–165. Ablex.

Palmer, Judith, and Simon Harding. 1992. Research reports: Can information users be classified like books? *Library and Information Research News* 15(54): 12–16.

Palmer, Judith. 1991. Scientists and information, II: Personal factors in information behaviour. *Journal of Documentation* 47(3): 254–275.

Paolillo, J. G. P. 1982. Technological gatekeepers: A managerial perspective. *IEEE Trans. on Engineering Management* EM-29: 169–171.

Paradis, James G., and Muriel L. Zimmerman. 1997. *The MIT Guide to Science and Engineering Communication.* Cambridge, MA: MIT Press.

Parker, E. B., W. J. Paisley, and R. Garrett. 1967. *Bibliographic Citations as Unobtrusive Measures of Scientific Communication.* Institute for Communication Research, Stanford University, San Francisco.

Parten, M. 1950. *Surveys, Pools and Samples: Practical Procedures.* Harper & Row and John Weatherhill.

Passman, Sidney. 1969. *Scientific and Technological Communication.* Oxford, England: Pergamon Press.

Patterson, M. 1968. Spatial factors in social interactions. *Human Relations* 21: 351–361.

Pauley, William. 1993. Knowledge utilization: The role of new communications technologies. *Journal of the American Society of Information Science* 44(4): 222–234.

Payne, S. L. 1973. *The Art of Asking Questions.* Princeton UP.

Pejtersen, Annelise Mark, Diane H. Sonnenwald, Jacob Buur, T. Govindaraj, and Kim Vicente. 1997. The Design Explorer project: Using a cognitive framework to support knowledge exploration. *Journal of Engineering Design* 8(3): 289–301.

Pelz, D. C., and F. M. Andrews. 1966. *Scientists in Organizations: Productive Climates for Research and Development.* New York: Wiley.

Penniman, W. David, David M. Liston, Jr., and Martin M. Cummings. 1992. *Final Report of the Conference for Exploration of a National Engineering Information Service.* Co-sponsored by Engineering Foundation and the Council on Library Resources. Ithaca, NY: Cornell Information Technologies and Media Services Printing.

Persson, O. 1980. Critical comments on the gatekeeper concept in science and technology. *RIT.* TRITA-LIB-6017.

Peterson, Ivars. 1990. The electronic grapevine: Computer networks and fax machines accelerate the pace of scientific communication—for good or ill. *Science News* 138(8): 90–91.

Pinelli, Thomas E., Rebecca O. Barclay, John M. Kennedy, and Ann Peterson Bishop. 1997a. *Knowledge Diffusion in the U.S. Aerospace Industry,* Part A. Greenwich, CT: Ablex.

Pinelli, Thomas E., Rebecca O. Barclay, John M. Kennedy, and Ann Peterson Bishop. 1997b. *Knowledge Diffusion in the U.S. Aerospace Industry,* Part B. Greenwich, CT: Ablex.

Note: Much of the work summarized in Pinelli, et al., 1997a and b is presented in more detail in the following reports and papers by Pinelli and co-authors.

Pinelli, Thomas E. 1992. Establishing a research agenda for scientific and technical information (STI): Focus on the user, paper 22. Paper presented at the *Research Agenda in Information Science* workshop sponsored by the Advisory Group for Aerospace Research and Development (AGARD), April 7–9, 1992, Lisbon, Portugal. Available from NTIS 92N28117.

Pinelli, Thomas E. 1991. The relationship between the use of U.S. government technical reports by U.S. aerospace engineers and scientists and selected institutional and sociometric variables, report 6. Washington, DC: National Aeronatics and Space Administration. NASA TM-102774. Available from NTIS 91N18898.

Pinelli, Thomas E. The information-seeking habits and practices of engineers, paper 13. Reprinted from *Science & Technology Libraries* 11(3): 5–25. Available from NTIS 92N28114.

Pinelli, Thomas E., and Rebecca O. Barclay. 1998. Maximizing the results of federally funded research and development through knowledge management: A strategic imperative for improving U.S. competitiveness. *Government Information Quarterly* 15(2): 157–172.

Pinelli, Thomas E., Rebecca O. Barclay, Ann P. Bishop, and John M. Kennedy. 1992. Information technology and aerospace knowledge diffusion: exploring the intermediary–end user interface in a policy framework, paper 23. Reprinted from *Electronic Networking: Research, Applications and Policy* 2(2): 31–49. (AIAA pending.)

Pinelli, Thomas E., Rebecca O. Barclay, Myron Glassman, and Walter E. Oliu. 1990. The value of scientific and technical information (STI), its relationship to research and development (R&D), and its use by U.S. aerospace engineers and scientists, paper 1. Paper presented at the *European Forum "External Information: A Decision Tool,"* January 19, 1990, Strasbourg, France. Available from AIAA 90A21931.

Pinelli, Thomas E., Rebecca O. Barclay, Stan Hannah, Barbara Lawrence, and John M. Kennedy. 1992. Knowledge diffusion and U.S. government technology policy: Issues and opportunities for sci/tech librarians, paper 27. Reprinted from *Science and Technology Libraries* 13(1): 33–55. Available from NTIS 93N20110 and A9329922.

Pinelli, Thomas E., Rebecca O. Barclay, Maurita P. Holland, Michael L. Keene, and John M. Kennedy. 1991. Technological innovation and technical communications: Their place in aerospace engineering curricula. A survey of European, Japanese and U.S. aerospace engineers and scientists, paper 21. Reprinted from the *European Journal of Engineering Education* 16(4): 337–351. Available from NTIS 92N28184.

Pinelli, Thomas E., Rebecca O. Barclay, Michael L. Keene, Madelyn Flammia, and John M. Kennedy. 1993. The technical communication practices of Russian and U.S. aerospace engineers and scientists, paper 28. Reprinted from *IEEE Transactions on Professional Communication* 36(2): 95–104. Available from NTIS N9328948.

Pinelli, Thomas E., Rebecca O. Barclay, Michael L. Keene, John M. Kennedy, and L. F. Hecht. 1995. From student to entry-level professional: Examining the role of language and written communication in the reacculturation of aerospace engineering students. *Technical Communication* 42(3): 492–503.

Pinelli, Thomas E., Rebecca O. Barclay, and John M. Kennedy. 1994. The U.S. government technical report and the transfer of federally funded aerospace R&D: An analysis of five studies, report 19. Washington, DC: National Aeronautics and Space Adminstration. NASA TM-109061, January. Available from NTIS 94N 24660.

Pinelli, Thomas E., Rebecca O. Barclay, and John M. Kennedy. 1994. The technical communication practices of British aerospace engineers and scientists: Results of the phase 4 RA3S mail survey, report 25. Washington, DC: National Aeronautics and Space Administration. NASA TM-109098, May. Available from NTIS 94N 32836.

Pinelli, Thomas E., Rebecca O. Barclay, and John M. Kennedy. 1993. The U.S. government technical report and the transfer of federally funded aerospace R&D. *Government Publications Review* 20(4): 393–411.

Pinelli, Thomas E., Rebecca O. Barclay, and John M. Kennedy. 1993. The U.S.

government technical report and the transfer of federally funded aerospace R&D, paper 29. Reprinted from *Government Publications Review* 20(3): 393–411. Available from NTIS 93N29709.

Pinelli, Thomas E., Rebecca O. Barclay, and John M. Kennedy. 1993. The relationship between technology policy and scientific and technical information within the U.S. and Japanese aerospace industries, paper 26. Paper presented at the *3rd Annual JICST/NTIS Conference on How to Locate and Acquire Japanese Scientific and Technical Information,* San Francisco, March 18. Available from NTIS 93N20111.

Pinelli, Thomas E., Rebecca O. Barclay, and John M. Kennedy. 1993. Users and uses of DOD technical reports: A report from the field, paper 34. Paper presented at the 1993 *Military Librarians Workshop,* Albuquerque, NM, November 15–19. (NTIS pending.)

Pinelli, Thomas E., Rebecca O. Barclay, and John M. Kennedy. 1993. A comparison of the technical communication practices of aerospace engineers and scientists in India and the United States, report 18. Washington, DC: National Aeronautics and Space Adminstration. NASA TM-109006, September, 68 pages. Available from NTIS 94N 13057.

Pinelli, Thomas E., Rebecca O. Barclay, and John M. Kennedy. 1994. The technical communication practices of U.S. aerospace engineers and scientists: Results of the Phase 1 SAE mail survey, report 24. Washington, DC: National Aeronautics and Space Administration. NASA TM-109088, May. Available from NTIS 94N 32837.)

Pinelli, Thomas E., Rebecca O. Barclay, and John M. Kennedy. 1994. The communications practices of U.S. aerospace engineering faculty and students: Results of the Phase 3 survey, report 23. Washington, DC: National Aeronautics and Space Administration. NASA TM-109085, April. Available from NTIS 94N 30149.

Pinelli, Thomas E., Rebecca O. Barclay, and John M. Kennedy. 1994. U.S. aerospace industry librarians and technical information specialists as information intermediaries: Results of the Phase 3 survey, report 22. Washington, DC: National Aeronautics and Space Adminstration. NASA TM-109067, March. Available from NTIS 94N 30150.

Pinelli, Thomas E., Rebecca O. Barclay, and John M. Kennedy. 1994. U.S. aerospace industry librarians and technical information specialists as information intermediaries: Results of the Phase 2 survey, report 21. Washington, DC: National Aeronautics and Space Administration. NASA TM-109064, February. Available from NTIS 94N 24709.

Pinelli, Thomas E., Rebecca O. Barclay, and John M. Kennedy. 1994. The use of selected information products and services by U.S. aerospace engineers and scientists: Results of two studies, report 20. Washington, DC: National Aeronautics and Space Adminstration. NASA TM-109022, February, 61 pages. Available from NTIS 94N 24649.

Pinelli, Thomas E., Rebecca O. Barclay, and John M. Kennedy. 1997. Survey of reader preferences concerning the format of NASA Langley-authored technical reports. *Publishing Research Quarterly* 13(2): 48–68.

Pinelli, Thomas E., Rebecca O. Barclay, John M. Kennedy, and Myron Glassman. 1990. Technical communications in aerospace: An analysis of the practices reported by U.S. and European aerospace engineers and scientists, paper 4. Paper presented at the *International Professional Communication Conference (IPCC)*, September 14, Guilford, England. Available from NTIS 91N14079 and AIAA 91A19799.

Pinelli, Thomas E., Ann Peterson Bishop, Rebecca O. Barclay, and John M. Kennedy. 1992. The electronic transfer of information and aerospace knowledge diffusion. *International Forum on Information and Documentation* 17(4): 8–16.

Pinelli, Thomas E., Ann P. Bishop, Rebecca O. Barclay, and John M. Kennedy. 1993. The information-seeking behavior of engineers, paper 31. Reprinted from the *Encyclopedia of Library and Information Science* 52, supplement 15: 167–201. Available from NTIS 93N30037.

Pinelli, Thomas E., and Myron Glassman. 1989. An evaluation of selected NASA scientific and technical information products: Results of a pilot study. Washington, DC: National Aeronautics and Space Administration. NASA TM-101533.

Pinelli, Thomas, E., Myron Glassman, Rebecca O. Barclay, and Walter E. Oliu. 1989. Technical communication in aerospace: Results of a Phase 1 pilot study—an analysis of managers' and nonmanagers' responses, report 2. Washington, DC: National Aeronautics and Space Administration. NASA TM-101625. Available from NTIS 90N11647.

Pinelli, Thomas E., Myron Glassman, and Vriginia M. Cordle. 1982. Survey of reader preferences concerning NASA technical reports. Washington, DC: National Aeronautics and Space Administration. NASA TM-84502. Available from NTIS 82N34300.

Pinelli, Thomas E., Myron Glassman, and Edward M. Gross. 1981. A review and evaluation of the Langley Research Center's Scientific and Technical Information Program: Results of Phase 1. Knowledge and Attitudes Survey, LaRC Research Personnel. Washington, DC: National Aeronautics and Space Administration. NASA TM-81893. Available from ERIC ED211052.

Pinelli, Thomas E., Myron Glassman, Walter E. Oliu, and Rebecca O. Barclay. 1989. Technical communications in aeronautics: Results of Phase 1 pilot study, report 1 (parts 1 and 2). Washington, DC: National Aeronautics and Space Administration. NASA TM-101534. February 1989. Part 1, available from NTIS 89N26772; Part 2, available from NTIS 89N26773.

Pinelli, Thomas, E., Myron Glassman, Walter E. Oliu, and Rebecca O. Barclay. 1989. Technical communications in aerospace: Results of Phase 1 pilot study—an analysis of profit managers' and nonprofit managers' responses, report 3. Washington, DC: National Aeronautics and Space Administration. NASA TM-101626, October 1989. Available from NTIS 90N15848.

Pinelli, Thomas E., and Nanci A. Glassman. 1992. Source selection and information use by U.S. aerospace engineers and scientists: Results of a telephone survey, report 13. Washington, DC: Nautional Aeronautics and Space Administration. NASA TM-107658.

Pinelli, Thomas E., Nanci A. Glassman, Linda O. Affelder, Laura M. Hecht, John M. Kennedy, and Rebecca O. Barclay. 1994. Technical uncertainty as a correlate of information use by U.S. industry-affiliated aerospace engineers and scientists, paper 36. Paper presented at the *32nd Aerospace Sciences Meeting of the American Institute of Aeronautics and Astronautics (AIAA)*, Reno, NV, January 1994. Available from AIAA 94-0839.

Pinelli, Thomas E., Madeline Henderson, Ann P. Bishop, and Philip Doty. 1992. Chronology of selected literature, reports, policy instruments, and significant events affecting federal scientific and technical information (STI) in the United States, report 11. Washington, DC: National Aeronautics and Space Administration. NASA TM-101662, January 1992. Available from NTIS 92N17001.

Pinelli, Thomas E., A. Rahman Khan, Rebecca O. Barclay, and John M. Kennedy. 1993. The U.S. government technical report and aerospace knowledge diffusion: Results of an ongoing investigation, paper 35. Paper presented at the *1st International Conference Grey Literature*, Amsterdam, The Netherlands, December 13–15, 1993. Available from AIAA.

Pinelli, Thomas E., and John M. Kennedy. 1991. The voice of the user: How U.S. Aerospace engineers and scientists view DOD technical reports, paper 11. Paper presented at the *Defense Tactical Information Center's (DTIC) Managers Planning Conference*, May 1, 1991, Solomon's Island, MD. Available from AIAA 91A41123.

Pinelli, Thomas E., and John M. Kennedy. 1990. Aerospace knowledge diffusion in the academic community: A report of Phase 3 activities of the NASA/DOD Aerospace Knowledge Diffusion Research Project, paper 6. Paper presented at the *Annual Conference of the American Society for Engineering Education—Engineering Libraries Division*, June 27, 1990, Toronto, Canada. Available from AIAA 91A19803.

Pinelli, Thomas E., and John M. Kennedy. 1990. Aerospace librarians and technical information specialists as information intermediaries: A report of Phase 2 activities of the NASA/DOD Aerospace Knowledge Diffusion Research Project, paper 5. Paper presented at the *Special Libraries Association, Aerospace Division—81st Annual Conference*, June 13, 1990, Pittsburgh, PA. Available from AIAA 91A19804.

Pinelli, Thomas E., and John M. Kennedy. 1990. The NASA/DOD Aerospace Knowledge Diffusion Research Project: The DOD perspective, paper 7. Paper presented at the *Defense Technical Information Center (DTIC) 1990 Annual Users Training Conference*, November 1, 1990, Alexandria, VA. Available from AIAA 91N28033.

Pinelli, Thomas E., John M. Kennedy, and Rebecca O. Barclay. 1993. A comparison of the technical communications practices of Russian and U.S. aerospace engineers and scientists, report 16. Washington, DC: National Aeronautics and Space Adminstration. NASA TM-107714, January 1993. Available from NTIS 93N18160.

Pinelli, Thomas E., John M. Kennedy, and Rebecca O. Barclay. 1991. The diffusion of federally funded aerospace research and development (R&D) and the information-seeking behavior of U.S. aerospace engineers and scientists, paper 12. Paper presented at the *Special Libraries Association (SLA) 92nd Annual Conference*, June 11, 1991, San Antonio, TX. Available from AIAA 92A29652.

Pinelli, Thomas E., John M. Kennedy, and Rebecca O. Barclay. 1991. The NASA/DOD Aerospace Knowledge Diffusion Research Project, paper 10. Reprinted from *Government Information Quarterly* 8(2): 219–233. Available from AIAA 91A35455.

Pinelli, Thomas E., John M. Kennedy, and Rebecca O. Barclay. 1990. The role of the information intermediary in the diffusion of aerospace knowledge, paper 8. Reprinted from *Science and Technology Libraries* 11(2): 59–76. Available from NTIS 92N28113.

Pinelli, Thomas E., John M. Kennedy, and Rebecca O. Barclay. 1995. Workplace communications skills and the value of communications and information use skills instruction: Engineering students' perspectives. In *Proceedings of the 1995 IEEE International Professional Communication Conference*, pp. 161–165. Piscataway, NJ: IEEE.

Pinelli, Thomas E., John M. Kennedy, Rebecca O. Barclay, Nanci A. Glassman, and Loren Demerath. 1991. The relationship between seven variables and the use of U.S. government technical reports by U.S. aerospace engineers and scientists, paper 17. Paper presented at the *54th Annual Meeting of the American Society for Information Science (ASIS)*, October 30, 1991, Washington, DC. Available from NTIS 92N28115.

Pinelli, Thomas E., John M. Kennedy, Rebecca O. Barclay, and Ann P. Bishop. 1992. Computer and information technology and aerospace knowledge diffusion, paper 19. Paper presented at the *Annual Meeting of the American Association for the Advancement of Science (AAAS)*, February 8, 1992, Chicago, IL. Available from NTIS 92N28211.

Pinelli, Thomas E., John M. Kennedy, Rebecca O. Barclay, and Terry F. White. 1991. Aerospace knowledge diffusion research, paper 16. Reprinted from *World Aerospace Technology '91: The International Review of Aerospace Design and Development* 1 (1991): 31–34. Available from NTIS 92N28220.

Pinelli, Thomas E., John M. Kennedy, and Terry F. White. 1991. Summary report to Phase 1 respondents, report 4. Washington, DC: National Aeronautics and Space Administration. NASA TM-102772, January 1991. Available from NTIS 91N17835.

Pinelli, Thomas E., John M. Kennedy, and Terry F. White. 1991. Summary report to Phase 1 respondents including frequency distributions, report 5. Washington, DC: National Aeronautics and Space Administration. NASA TM-102773. Available from NTIS 91N20988.

Pinelli, Thomas E., John M. Kennedy, and Terry F. White. 1991. Summary report to Phase 2 respondents including frequency distributions, report 7. Washington, DC: National Aeronautics and Space Administration. NASA TM-104063, March 1991. Available from NTIS 91N22931.

Pinelli, Thomas E., John M. Kennedy, and Terry F. White. 1991. Summary report to Phase 3 academic library respondents including frequency distributions, report 10. Washington, DC: National Aeronautics and Space Administration. NASA TM-104095, August 1991. Available from NTIS 91N33013.

Pinelli, Thomas E., John M. Kennedy, and Terry F. White. 1991. Summary report to Phase 3 faculty and student respondents including frequency distributions, report 9. Washington, DC: National Aeronautics and Space Administration. NASA TM-104086, June 1991, 42. Available from NTIS 91N25950.

Pinelli, Thomas E., John M. Kennedy, and Terry F. White. 1991. Summary report to Phase 3 faculty and student respondents, report 8. Washington, DC: National Aeronautics and Space Administration. NASA TM-104085, June 1991. Available from NTIS 91N24943.

Pinelli, Thomas E., John M. Kennedy, and Terry F. White. 1992. Engineering work and information use in aerospace: Results of a telephone survey, report 14. Washington, DC: National Aeronautics and Space Administration. NASA TM-107673, October 1992. (NTIS pending.)

Pinelli, Thomas E., Yuko Sato, Rebecca O. Barclay, and John M. Kennedy. 1997d. Culture and workplace communications: A comparison of the technical communications practices of Japanese and U.S. aerospace engineers and scientists. *Journal of Air Transportation World Wide* 2(1): 1–21.

Pinelli, T. E., et al. 1980. A review and evaluation of the Langley Research Center's Scientific and Technical Information Program. Results of Phase 1— knowledge and attitudes survey. LARC Research Personnel. NASA-TM-81893. NASA.

Plumb, Carolyn, and Cathie Scott. 2000. A successful process for developing performance-based outcomes for engineering student writing assessment. In *Proceedings of the 2000 Frontiers in Education Conference,* T4A–18. Piscataway, NJ: IEEE.

Poland, Jean. 1993. Informal communication among scientists and engineers: A review of the literature. *Science and Technology Libraries* 11(3): 61–73.

Poli, Corrado. 1996. Engineering communication skills and design for manufacturing: A freshman engineering course. In *Proceedings of the SME International Conference on Education in Manufacturing.* Available from World Wide Web: <http://www.ecs.umass. edu/mie/faculty/poli/freshman.html>.

Poole, Herbert L. 1979. Information use: A synthesis of existing knowledge via theories of the middle range Ph.D. diss., Rutgers University.

Poppel, H. L. 1980. Managerial/professional productivity. Booze, Allen and Hamilton.

Posey, Edwin D., and Charlotte A. Erdman. 1986. An online UNIXÆ-based engineering catalog: Purdue University Engineering Library. In *Role of Computers in Sci-Tech Libraries,* pp. 31–43. New York: Haworth Press.

Prausnitz, Mark R., and Melissa J. Bradley. 2000. Effective communication for professional engineering: Beyond problem sets and lab reports. *Chemical Engineering Education* 34(3): 234–237.

Pries, Frens, and Felix Janszen. 1995. Innovation in the construction industry: The dominant role of the environment. *Construction Management and Economics* 13(1): 43–51.

Pryor, H. E. 1975. An evaluation of the NASA Scientific and Technical Information System. *Special Libraries* 66: 515–519.

Pryor, H. E. 1976. Listening to the user: A case study. In *The Problem of Optimization of User Benefit in Scientific and Technological Information Transfer. AGARD.* AGARD-CP-179: 11.1–11.7.

Pullinger, David. 1996. SuperJournal: A project in the UK to develop multimedia journals. *D-Lib Magazine.* Available from World Wide Web: <http://www.dlib.org/dlib/january96/briefings/01super.html>.

Pullinger, David J., and Christine Baldwin. 2002. *Electronic Journals and User Behaviour: Learning for the Future from the SuperJournal Project.* Cambridge, England: deedot Press.

Quinn, John J. 1985. Information and the industrial chemist. *Chemistry in Britain* 21(8): 738–739.

Quiroz, Sharon. 1999. Review: Special issue on engineering communication. In *Proceedings of the 1999 Frontiers in Education Conference,* 13b5–6. Piscataway, NJ: IEEE.

Raitt, David Iain. 1986. Introducing information systems to scientists and engineers. *South African Journal of Library and Information Science* 54: 97–101.

Raitt, David Iain. 1985. The information-seeking and communication habits of scientists and engineers. *ASIS Proceedings,* pp. 319–323.

Raitt, David Iain. 1984. The communication and information-seeking and use habits of scientists and engineers in international organizations based in Europe national aerospace research establishments. Ph.D. diss., Loughborough University of Technology.

Raitt, D. 1980a. Marketing and promotion of online services for intermediaries. In *Proceedings of 4th International Online Information Meeting,* December 9–11, 1980, pp. 265–274. London, England: Learned Information.

Raitt, D. 1980b. Teaching the teachers: Essentials of online educating for intermediaries. In *Proceedings of 4th International Online Information Meeting,* December 9–11, 1980, London, England. Learned Information. pp. 427–434.

Raitt, D., 1982a. Recent developments in telecommunications and their impact on information services. *ASLIB Proceedings* 34: 54–76.

Raitt, D., 1982b. The information specialist: Transition to information manager. In B. Hubbard, ed., pp. 249–260. State of the art report. *Technology Management.* Pergamon-Infotech.

Raitt, D. 1983. Information technology and the library. *The Electronics Library* 1: 149–156.

Ramstrom, D. 1967. *The Efficiency of Control Strategies: Communications and Decision-Making in Organizations.* Almqvist and Wiksell.

Randolph, Gary B. 2000. Collaborative learning in the classroom: A writing across the curriculum approach. *Journal of Engineering Education* 89: 119–122.

Rawdin, Eugene. 1975. Field survey of information needs of industry sci/tech library users. In *Proceedings of 38th ASIS Annual Meeting,* Boston, *ASIS,* pp. 41–42.

Rawdin, Eugene. 1975. Field survey of information needs of industry sci/tech users. In *Information Revolution: 38th Annual ASIS Meeting.* Charles W. Husbands, ed., pp. 147–148. Washington, DC: American Society for Information Science.

Read, W. H. 1962. Upward communication in industrial hierarchies. *Human Relations* 15: 3–15.

Redfield, C. E. 1954. *Communication in Management.* Chicago: University of Chicago Press.

Rees-Jones, Lyndsay. 1998. Snapshot: The motor engineering industry—all revved up and ready to go. *Library Association Record* 100(6): 304–305.

Report of the Comptroller General of the United States. 1976. Observations on collection and dissemination of scientific, technical, and engineering information. Washington, DC: National Technical Information Service, U.S. General Accounting Office.

Resh, Vincent H. 1998. Science and communication: An author/editor/user's perspective on the transition from paper to electronic publishing. *Issues in Science and Technology Librarianship* 19(Summer). Available from World Wide Web: <http://www.library.ucsb. edu/istl/98-summer/article3.html>.

Rice, R. E., and J. Bair. 1983. Conceptual role of new communication technology in organizational productivity. In *Proceedings of 46th ASIS Annual Meeting,* October 2–6, Washington, DC, *ASIS,* Vol. 20: 4–8.

Richards, J. W. 1967. *Interpretation of Technical Data.* Van Nostrand.

Richardson, Robert J. 1981. End-user online searching in a high-technology engineering environment. *Online* 5(4): 44–57.

Rickards, Janice, Peter Linn, and Diana Best. 1989. Information needs and resources of engineering firms: Survey of Brisbane and the Gold Coast of Queensland. *Australasian College Libraries* 7: 63–72.

Rieh, Hae-young. 1993. Citation analysis: A case study of Korean scientists and engineers in electrical and electronics engineering. In *56th American Society for Information Science Annual Meeting,* pp. 165–171.

Ritchie E., and A. Hindle. 1976. *Communication Networks in R&D: A Contribution to Methodology and Some Results in a Particular Laboratory.* University of Lancaster, Department of Operational Research. BLRDR 5291.

Robar, Tracy Y. 1998. Communication and career advancement. *Journal of Management in Engineering* 14(2): 26–28.

Robbins, J. C. 1973. Social functions of scientific communication. *IEEE Trans. on Professional Communication* PC-16: 131–135, 181.

Roberts, A. H. 1970. The system of communication in the language sciences: Present and future. In *Communication among Scientists and Engineers.* C. E Nelson and D. K Pollock, eds., pp. 293–306. Heath Lexington Books.

Robertson, A. 1973. Information flow and industrial innovation. *ASLIB Proceedings* 25: 130–139.

Robertson, I. T., and C. L. Cooper. 1983. *Human Behavior in Organizations.* Macdonald and Evans.

Robinson, John. 2000. "New model" engineers take top jobs in high-tech sector. *The Engineer* (January 28): 35.

Robinson, Joseph A. 1997. Communication and engineers: Collisions at the crossroads! In *Proceedings of the 1997 IEEE International Professional Communication Conference,* pp. 419–426. Piscataway, NJ: IEEE.

Roderer, Nancy K., and Donald W. King. 1982. Information dissemination and technology transfer in telecommunications. November 1982. Available from ERIC ED 239582.

Rodriguez, Martius V. R., Norberto Cotta Siamo, Celi Lorenzoni, and A. J. Ferrante. 1994. Processing and distribution of graphical information in offshore engineering. In *Proceedings of the 4th International Offshore and Polar Engineering Conference,* pp. 177–180. Golden, CO: International Society of Offshore and Polar Engineers.

Rogers, Everett M. 1983. *Diffusion of Innovations.* New York: Free Press.

Rogers, Everett M. 1982. Information exchange and technological innovation. In *The Transfer and Utilization of Technical Knowledge.* Devendra Sahal, ed., pp. 49–60. Lexington, MA: D.C. Heath.

Rogers, E. M., and R. Agarwala-Rogers. 1976. *Communication in Organizations.* New York: Free Press.

Rogers, E. M., and R. Agarwala-Rogers. 1980. Three schools of organizational behavior. In *Intercom: Readings in Organizational Communication.* S. Ferguson and S. Devreaux Ferguson, eds., pp. 2–31. Hayden.

Rogers, E. M., and F. F. Shoemaker. 1971. *Communication of Innovations,* 2nd ed. New York: Free Press.

Rogers, Sally A. 2001. Electronic journal usage at Ohio State University. *College & Research Libraries* 62(1): 25–34.

Rose-Hulman Institute of Technology. 2000. Available from World Wide Web: <http://www.rose-hulman.edu/IRA/IRA/index.html>.

Rosenberg, Victor. 1967. Factors affecting the preferences of industrial personnel for information gathering methods. *Information Storage and Retrieval* 3: 119–127.

Rosenbloom, Richard S., and Francis W. Wolek. 1967. *Technology, Information, & Organization: Information Transfer in Industrial R&D.* Boston: Harvard University.

Rosenbloom, Richard S., and Francis W. Wolek. 1970. *Technology and Information Transfer: A Survey of Practice in Industrial Organizations.* Boston: Division of Research, Graduate School of Business Administration, Harvard University.

Rostron, S. 1979. Tomorrow's office. *Reprographic Quarterly* 13(4): 135–138.

Rothwell, Roy (University of Sussex). 1975. Patterns of information flow during the innovation process. *ASLIB Proceedings* 27(5).

Rothwell, R., and A. B. Robertson. 1973. The role of communication in technological innovation. *Research Policy* 2: 204–225.

Rothwell, R., and A. B. Robertson. 1975. The contribution of poor communication to innovative failure. *ASLIB Proceedings* 27(10): 393–400.

Rowley, J. E., and C. M. D. Turner. 1978. *The Dissemination of Information.* Boulder, CO: Westview Press.

Rubenstein, A., et al. 1970. Explorations on the information-seeking style of researchers. In *Communication Among Scientists and Engineers.* C. E. Nelson and D. K. Pollock, eds., pp. 209–231. Heath Lexington Books.

Rubenstein, A. H., et al. 1971. Ways to improve communications between R&D groups. *Research Management* 14(6): 49–59.

Rubenstein, Albert H., C. W. N. Thompson, and Robert D. O'Keefe. 1976. Critical field experiments on uses of scientific and technical information. *Current Research on Scientific and Technological Information Transfer.* Micropapers edition. New York: Jeffery Norton Publishers.

Rusch-Feja, Diann, and Uta Siebeky. 1999. Evaluation of usage and acceptance of electronic journals: Results of an electronic survey of Max Planck Society researchers including usage statistics from Elsevier, Springer and Academic Press. *D-Lib Magazine* 5(10). Available from World Wide Web: <http://www. dlib. org/dlib/october99/rusch-feja/10rusch-feja-summary. html>.

Rzevski, G. 1985. On criteria for assessing an information theory. *Computer Journal* 28(3): 200–202.

Sabharwal, Kapil, and A. J. Nicholson. 1997. Internet communications technologies and their effect on the environmental profession. *Practice Periodical of Hazardous, Toxic, and Radioactive Waste Management* 1(3): 120–123.

Sabine, Gordon A., and Patricia L. Sabine. 1986. How people use books and journals. *Library Quarterly* 56(4): 399–408.

Saha, J. 1970. User studies for evaluation of information systems and library resources. In *Proceedings of the 35th FID Conference and International Congress of Documentation*. Buenos Aires, Argentina: FID.

Salzberg, S., and M. Watkins. 1990. Managing information for concurrent engineering: Challenges and barriers. *Research in Engineering Design* 2: 35–52.

Sampson, A. 1968. *The New Europeans.* Hodder and Stoughton.

Saracevik, Tefko. 1970. Ten years of relevance assessments: A summary and synthesis of conclusions. In *ASIS 33rd Annual Meeting Proceedings,* pp. 33–36.

Saracevic, Tefko, Paul Kantor, A. Y. Chamic, and D. Trivison. 1988. A study of information seeking and retrieving, Parts I, II, III. *Journal of the American Society for Information Science* 39(3): 162–216.

Schachter, S. 1968. Deviation, rejection and communication. In *Group Dynamics: Research and Theory,* 3rd ed. D. Cartwright, and A. Zander, eds., pp. 165–181. London: Tavistock Pubs.

Schaefermeyer, Mark J., and Edward H. Sewell, Jr. 1988. Communicating by electronic mail. *American Behavioral Scientist* 32(2): 112–123.

Scharf, Davida. [See Council on Library Resources, 1993].

Schauder, Don. 1994. Electronic publishing of professional articles: Attitudes of academics and implications for the scholarly communications industry. *Journal of the American Society for Information Science* 45(2): 73–100.

Schein, E. H. 1961. Interpersonal communication, group solidarity and social influence. In *The Planning of Change.* W. G. Bennis, et al., eds., pp. 517–527. Holt, Rinehart and Winston.

Schillaci, William C. 1996. Training engineers to write: Old assumptions and new directions. *Journal of Technical Writing and Communication* 26(3): 325–333.

Schlesinger, L. E. 1972. Meeting the risks involved in two-way communications. *Personnel Administration* 25(6): 24–30.

Schmidt, J. 1980. The nature and importance of user education and its relevance to the special library situation. *Australian Special Libraries News* 31(1): 9–13.

Schrage, Michael, and Alun Anderson. 1991. Computer tools for thinking in tandem: "Groupware" can erase geography. It may supplant printed journals and link researchers in "virtual laboratories." *Science* 253(5019) (August 2, 1991): 505–507.

Schuler, R. S., and L. F. Blank. 1976. Relationships among types of communication, organizational level and employee satisfaction and performance. *IEEE Transactions on Engineering Management* EM-23(3): 124–129.

Schum, David A. 2000. Teaching about discovery and invention in engineering. *Technological Forecasting and Social Change* 64(2/3): 209–223.

Scott, Bill. 1984. *Communication for Professional Engineers*. London: Thomas Telford Ltd.

Scott, Christopher. 1959. The use of technical literature by industrial technologists. In *Proceedings of the International Conference on Scientific Information*, pp. 245–266. Washington, DC: NAS.

Scott, Christopher. 1962. The use of technical literature by industrial technologists. *IRE Transactions on Engineering Management*, EM-9(2): 76–86.

Scott, W. G. 1969. Organization theory: An overview and appraisal. In *Readings in Management*, 3rd ed., M. D. Richards and W. A. Nielander, eds., pp. 664–688. South-Western Publishing Company.

Selznick, P. 1971. Foundations of the theory of organizations. In *Systems Thinking: Selected Readings*. F. E. Emery, ed., pp. 261–280. Penguin.

Senders, J. W. 1970. Some thoughts on scientific communication or who does what with which and to whom? In *Innovations in Communication Conference*, PB 192294-13, CFSTI-70-01, 184-188. CFSTI.

Senders, J. W., C. M. B. Anderson, and C. P. Hecht. 1975. Scientific publication systems: An analysis of past, present, and future methods of scientific communication. Toronto: University of Toronto, 1975. Available from NTIS PB 242259.

Shapero, Albert. 1976. The effective use of STI in industrial and non-profit settings: Exploration through experimental interventions in ongoing R&D activities. Austin, TX: University of Texas. Available through ERIC: ED 121 269.

Sharp, E. T. 1975. Applying the user/system interface analysis results to optimize information transfer. In *Information Revolution: 38th Annual ASIS Meeting*, Charles W. Husbands, ed. Washington, DC: American Society for Information Science.

Sharp, Julie E. 1998. Learning styles and technical communication: Improving communication and teamwork skills. In *Proceedings of the 1998 Frontiers in Education Conference*, pp. 512–517. Piscataway, NJ: IEEE.

Shelfer, Katherine M. 1998. Understanding science and technology research needs. *Florida Libraries* 41(6): 123–127.

Sheppard, Margaret O. 1975. User response to the SDI service developed at Aeronautical Research Laboratories, Australia. In *Information Revolution: 38th Annual ASIS Meeting*. Charles W. Husbands, ed. Washington, DC: American Society for Information Science.

Sheppard, Margaret O. 1976. User response to the SDI service developed at Aeronautical Research Laboratories, Australia. In *The Problem of Optimization of User Benefit in Scientific and Technological Information Transfer*, AGARD CP-179, 10-1–10-9. AGARD.

Shilling, C. W., J. Bernard, and J. Tyson. 1964. *Informal Communication among Bioscientists.* Biological Science Communication Project. Washington, DC: George Washington University.

Shoham, Snunith. 1998. Scholarly communication: A study of Israeli academic research. *Journal of Librarianship and Information Science* 30(2): 118–121.

Shotwell, Thomas K. 1971. Information flow in an industrial research laboratory: A case study. *IEEE Transactions on Engineering Management* 18(1): 26–33.

Shuchman, Hedvah L. 1983. Engineers who patent: Data from a recent survey of American bench engineers. *World Patent Information* 5(3): 174–179.

Shuchman, Hedvah L. 1982. Information technology and the technologist: A report on a national study of American engineers. *International Federation for Documentation* 7(1): 3–8.

Shuchman, Hedvah L. 1981. *Information Transfer in Engineering.* Glastonbury, CT: The Futures Group.

Shuchman, Hedvah L. 1980. Informal information networks and women in engineering. In *43rd ASIS Annual Meeting,* Anaheim, CA, October 5–10, 1980.

Siess, Judith A. 1982. Information needs and information-gathering behavior of research engineers. In *Proceedings of the 11th ASIS Midyear Meeting,* Knoxville, TN, June 13–16, 1982.

Sieving, Pamela C. 1991. The information quest as resolution of uncertainty: Some approaches to the problem. *Science and Technology Libraries* 11(3): 75–81.

Sih, Julie, and Christy Hightower. 1994. When in-person classes aren't the answer: Teaching INSPEC electronically. *DLA Bulletin* 14(1): 6–10.

Sim, Susan Elliott, Janice Singer, and Margaret Anne Storey. 2001. Beg, borrow, or steal: Using multidisciplinary approaches in empirical software engineering research. *Empirical Software Engineering* 6(1): 85–93.

Simons, G. R. 1994. Facilitating the role of the knowledge supplier in customer product design. In *Proceedings of the 4th International Conference on Flexible Automation and Integrated Manufacturing (FAIM '94),* pp. 303–310. Blacksburg, VA.

Sims, H. P., and W. LaFollette. 1975. An assessment of the Litwin and Stringer organization climate questionnaire. *Personnel Psychology* 28: 19–38.

Skelton, Barbara. 1971. Comparison of results of science and user studies with investigation into information requirements of the social sciences. Bath, UK: Bath University Library.

Skelton, B. 1973. Scientists and social scientists as information users: A comparison of results of science user studies with the investigation into information requirements of the social sciences. *Journal of Librarianship* 5(2): 138–156.

Skeris, Peter. 1998. Engineers becoming better managers: Overcoming communication barriers. In *ISA TECH/EXPO Technology Update Conference Proceedings* 2(3): 31–39.

Skinder, Robert F., and Robert S. Gresehover. 1995. An Internet navigation tool for the technical and scientific researcher. *Online* 19(4): 38–42.

Slater, M. 1981. The neglected resource: Non-usage of library-information services in industry and commerce. *BLR&D Report 5628.* London: ASLIB.

Smith, B. B. 1983. Marketing strategies for libraries. *Library Management* 4(1): 1–52.

Smith, Clagett G. 1970. Consultation and decision processes in a research and development laboratory. *Administrative Science Quarterly* 15(2): 203–215.

Smith, Elaine Davis. 1993. A comparison of the effects of new technology on searching habits in industrial and academic institutions. *Journal of Information Science* 19: 57–66.

Smith, F. R. 1977. Enhancing technical communications within a large corporation. *Technical Communication* 24(4), 4th Quarter: 12–16.

Solla Price, D. J. de. 1963. *Little Science, Big Science.* Columbia UP.

Sonnenwald, Diane H., and Linda G. Pierce. 2000. Information behavior in dynamic work group contexts: Interwoven situational awareness, dense social networks and contested collaboration in command and control. *Information Processing & Management* 36(3): 461–479.

Spencer, Richard H., and Raymond E. Floyd. 1995. Educating technical professionals to communicate. In *Proceedings of 1995 IEEE International Professional Communication Conference,* pp. 152–155. Piscataway, NJ: IEEE.

Spilka, Rachel. 1990. Orality and literacy in the workplace: Process- and text-based strategies for multiple-audience adaptation. *Journal of Business and Technical Communication: JBIC* 4(1): 44–67.

Spretnak, Charles M. 1982. A survey of the frequency and importance of technical communication in an engineering career. *Technical Writing Teacher* 9: 133–136.

Stenzler-Centonze, Marjorie. 1990. EE's next challenge: People, not products. (Electrical engineers must improve interpersonal skills.) *EDN* 35(17A): 39–40.

Sterling, Theodor D. 1988. Analysis and reanalysis of shared scientific data. *Annals of the American Academy of Political and Social Sciences,* pp. 49–60.

Stern, Arnold. 1989. Information transfer between an academic research center and its member firms. *Journal of Technology Transfer* 14(314): 19–24.

Sternberg, Virginia Ashworth. 1991. Use of federally supported IACs by special libraries in large companies. Ph.D. diss., University of Pittsburgh.

Strain, Paula M. 1973. Engineering libraries: A user survey. *Library Journal* 98(9): 1446–1448.

Studt, T. 1998. How researchers use the Internet. *R&D Magazine* 40(2): 20–27.

Subramanyam, K. 1981. *Scientific and Technical Information Sources.* New York: Marcel Dekker.

Summers, Edward G., Joyce Matheson, and Robert Conry. 1983. The effect of personal, professional and psychological attributes and information-seeking behavior on the use of information sources by educators. *Journal of the American Society for Information Science* 34(1): 75–85.

Sutton, J. R. 1975. Information requirements of engineering designers. In

Information Revolution: 38th Annual ASIS Meeting. Charles W. Husbands, ed., pp. 147–148. Washington, DC: American Society for Information Science.

Taylor, Robert L. 1977. A longitudinal analysis of technical communication in research and development. *Journal of Technology Transfer* 1(2): 17–31.

Taylor, Robert L. 1975. The technological gatekeeper. *R&D Management* 5(3): 239–242.

Taylor, Robert L., and James M. Utterback. 1975. A longitudinal study of communication in research: technical and managerial influences. *IEEE Transactions on Engineering Management,* EM-22, no. 2 (May): 80–87.

Taylor, R. S. 1986. *Value-added Processes in Information Systems.* Norwood, NJ: Ablex.

Taylor, Robert S. 1991. Information use environments. In *Progress in Communication Services,* 10th ed. Brenda Dervin and Melvin Voigt, eds., pp. 217–255. Norwood, NJ: Ablex.

Tedd, L. A. 1981. Teaching aids developed and used for education and training for online searching. *Online Review* 5(3): 205–216.

Tenopir, Carol. 1996. Moving to the Information Village. *Library Journal* 121 (March 2): 29–30.

Tenopir, Carol. 2003. *Use and Users of Electronic Library Resources: An Overview and Analysis of Recent Research Studies.* Washington D.C.: Council on Library and Information Resources. Available from World Wide Web: *<http://www.clir.org/pubs/reports/pub120/pub120.pdf>.*

Tenopir, Carol, D. W. King, P. Boyce, M. Grayson, Y. Zhang, and M. Ebuen. 2003. Patterns of journal use by scientists through three evolutionary phases. *D-Lib Magazine,* 9(5). Available from World Wide Web: *<doi:10.1045/may2003-king>.*

Tenopir, Carol, and Donald W. King. 2001. The use and value of scientific journals: Past, present, and future. *Serials* (July/August): 113–120.

Tenopir, Carol, and Donald W. King. 2000a. *Towards Electronic Journals: Realities for Scientists, Librarians, and Publishers.* Washington, DC: Special Libraries Association.

Tenopir, Carol, and Donald W. King. 2000b. The use and value of scholarly journals. In *Proceedings of the 63rd Annual Meeting of the American Society of Information Science* 37: 60–62. Chicago: ASIS.

Tenopir, Carol, and Donald W. King. 1998a. Designing the future of electronic journals with lessons learned from the past: Economic and use patterns of scientific journals. In *Proceedings of the Socioeconomic Dimensions of Electronic Publishing Workshop,* pp. 11–17. Santa Barbara, CA: 1998 IEEE Advances in Digital Libraries Conference.

Tenopir, Carol, and Donald W. King. 1998b. Designing the future of electronic journals with lessons learned from the past: Economic and use patterns of scientific journals. *Journal of Electronic Publishing.* Available from World Wide Web: <http://www.press.umich.edu/jep/04-02/king.html>.

Tenopir, Carol, and Donald W. King. 1997a. Managing scientific journals in a digital era. *Information Outlook* 1: 14–17.

Tenopir, Carol, and Donald W. King. 1997b. Trends in scientific scholarly publishing in the U.S. *Journal of Scholarly Publishing* 28: 135–170.

Tenopir, Carol, and Donald W. King. 1996. Electronic publishing: A study of functions and participants. In *17th National Online Meeting Proceedings,* pp. 375–384. Medford, NJ: Learned Information.

Tenopir, Carol, Donald W. King, Randy Hoffman, Elizabeth McSween, Christopher Ryland, and Erin Smith. 2001. Scientists' use of journals: Differences (and similarities) between print and electronic. In *Proceedings of the 22nd National Online Meeting,* pp. 469–481. Medford, NJ: Information Today.

Tharp, P. A. 1971. *Regional International Organization: Structures and Functions.* Macmillan.

Thomas, Rick, and Robert Drury. 1988. Team communication in complex projects. *Engineering Management International* 4: 287–297.

Thompson, Benna. 1982. Future direct users of sci-tech electronic bases. *Proceedings of the 11th ASIS Midyear Meeting,* Knoxville, TN, June 13–16, 1982.

Thompson, Charles W. N. 1975. Technology utilizations. In *Annual Review of Information Science and Technology* 10: 385–414. Washington, DC: American Society for Information Science.

Tocatlian, J. 1978. Training information users: Programmes, problems, prospects. *UNESCO Bulletin for Libraries* XXXII (6): 355–362.

Tombaugh, Jo W. 1984. Evaluation of an international scientific computer-based conference. *Journal of Social Issues* 40(3): 129–144.

Toraki, Katerina. 1999. Greek engineers and libraries in the coming years: A (human) communication model. *IATUL Proceedings* 9. Available from World Wide Web: <http://educate.lib.chalmers.se/IATUL/proceedcontents/chanpap/toraki.html>.

Torgerson, W. S. 1958. *Theory and Methods of Scaling.* New York: Wiley.

Tornudd, E. 1959. Study on the use of scientific literature and reference services by Scandinavian scientists and engineers engaged in research and development. In *Proceedings of the International Conference on Scientific Information,* Washington DC, November, 16–21, 1958, NAS, 1: 19–75.

Triandis, H. C. 1959. Cognitive similarity and interpersonal communication in industry. *Journal of Applied Psychology* 43(5): 321–326.

Triandis, H. C. 1960. Some determinants of interpersonal communication. *Human Relations* 13: 279–287.

Tucci, Valerie K. 1988. Information marketing for libraries. In *Annual Review of Information Science and Technology* 23. Amsterdam: Elsevier Science Publishers.

Tuck, Bill, Cliff McKnight, Marie Hayet, and David Archer. 1990. Project Quartet. *Library and Information Research Report* 76. London: British Library.

Turoff, M., and S. R. Hiltz. 1982. The electronic journal: A progress report. *Journal of the American Society for Information Science* 33(4).

Turoff, Murray, and Julian Scher. 1975. New Jersey institute of technology. Computerized conferencing and its impact on engineering management. *Joint Engineering Management Conference,* pp. 59–70.

Tushman, Michael L. 1976. *Communication in research and development organizations: An information processing approach.* Ph.D. diss., MIT.

Tushman, Michael L. 1977. Special boundary roles in the innovation process. *Administrative Science Quarterly* 22(4): 587–635.

Tushman, Michael L. 1978. Technical communication in R&D laboratories: The impact of project work characteristics. *Academy of Management Journal* 21(4): 624–645.

Tushman, Michael L. 1979. Impacts of perceived environmental variability on patterns of work related communication. *Academy of Management Journal* 22(3): 482–500.

Tushman, Michael L. 1981. Managing communication networks in R&D laboratories. *IEEE Engineering Management Review* 9(4): 65–77.

Tushman, Michael. 1979. Managing communication networks in R&D laboratories. *Sloan Management Review* 20 (Winter): 37–49.

Tushman, M. L., and R. Katz. 1980. External communication and project performance: An investigation into the role of gatekeepers. *Management Science* 26(11): 1071–1085.

Tushman, M. L., and D. A. Nadler. 1980. Communication and technical roles in R&D laboratories: An information processing approach. In *Management of Research and Innovation.* B. V. Dean and J. L. Goldhar, eds. North-Holland.

Tushman, Michael L., and David A. Nadler. 1978. Information processing as an integrating concept in organizational design. *Academy of Management Review* 3(1): 613–624.

Tushman, Michael L., and Thomas J. Scanlan. 1981. Boundary spanning individuals: Their role in information transfer and their antecedents. *Academy of Management Journal* 24(2): 289–305.

Ulrich's Guide to Periodicals. 2001. New Providence, NJ: R.R. Bowker.

UNESCO. 1976. *Final Report of UNISIST Seminar on the Education and Training of Users of Scientific and Technological Information* (Oct. 18–21, 1976, Rome, Italy). UNESCO, SC-77/WS/22.

UNESCO. 1981. *Guidelines on Studies of Information Users (Pilot Version).* UNESCO, PG1/81/WS/2.

Union of International Associations. 1983. *Yearbook of International Organizations 1983/84, Vol. 1.* Saur.

U.S. Congress. 1986. Office of Technology Assessment. Intellectual property rights in an age of electronics and information. Washington, DC: U.S. Government Printing Office, 1986.

U.S. Congress. 1989. Office of Technology Assessment. Staff paper. Federal scientific and technical information in an electronic age: Opportunities and challenges. Washington, DC: Office of Technology Assessment, p. 1.

U.S. Department of Commerce. 1988. Survey of supply/demand relationships for Japanese technical information in the U.S.: The field of advanced ceramics R&D. Available from NTIS: PB88-210943.

United States. 1986. House Committee on Science. Task Force on Science Policy. Science policy study, Background Report no. 5: The impact of information technology on science. Transmittal to the 99th Congress, 2nd session, September 1986. Prepared by Congressional Research Service, Library of Congress.

Utterback, James M. 1971. The process of innovation: A study of the origination and development of ideas for new scientific instruments. *IEEE Transactions on Engineering Management* EM-18(4): 124–131.

Van House, Nancy A., Beth T. Weil, and Charles R. McClure. 1990. *Measuring Academic Library Performance.* Chicago: American Library Association.

Van Styvendale, B. J. H. 1977. University scientists as seekers of information: Sources of references to periodical literature. *Journal of Librarianship* 9: 270–277.

Van Styvendale, B. J. H. 1981. University scientists as seekers of information. Sources of references to books and their first use versus date of publication. *Journal of Librarianship* 13: 83–92.

Vaughan, N. D., and M. J. Shipway. 1995. Educating British engineers to work in Europe. *European Journal of Engineering Education* 20(1): 75–81.

Veenhoff-Lovering, A. 1983. *Educating the Special Library User: An Annotated Bibliography 1960–1983.* Unpublished.

Vergeest, J. S. M., E. J. J. van Breeman, W. G. Knoop, and T. Wiegers. 1995. An effective method to analyze chronological information aspects in actual engineering processes. In *Proceedings of Computer Applications in Production and Engineering,* pp. 133–142. London: Chapman & Hall.

Veshosky, David. 1998. Managing innovation information in engineering and construction firms. *Journal of Management in Engineering* 14(1): 58–66.

Vest, David, Marilee Long, and Thad Anderson. 1996. Electrical engineers' perceptions of communication training and their recommendations for curricular change: Results of a national survey. *IEEE Transactions on Professional Communication* 39(1): 38–42.

Vest, David, Marilee Long, Laura Thomas, and Michael E. Palmquist. 1995. Relating communication training to workplace requirements: The perspective of new engineers. *IEEE Transactions on Professional Communication* 38(1): 11–17.

Veyette, Jr., John H., Robert Bezilla, and Y. S. Touloukian. 1978. Alternatives for accessing engineering numerical data. New York: Engineering Information. Available from NTIS PB 282609.

Vickery, B. C. 1973. *Information Systems.* Connecticut: Archon Books, pp. 33–52.

Vincenti, Walter G. 1990. *What Engineers Know and How They Know It.* Baltimore, MD: Johns Hopkins University Press.

Vincler, James E., and Nancy Horlick Vincler. 1997. Writing for success. *Chemical Engineering* 104(2): 111–113.

Von Seggern, Marilyn, and Janet M. Jourdain. 1996. Technical communication in engineering and science: The practices within a government defense laboratory. *Special Libraries* 87(2): 98–119.

Voos, H. 1967. *Organizational Communication: A Bibliography.* Rutgers UP.

Wagner, Michael M., and Gregory F. Cooper. 1992. Evaluation of Meta-1-based Automatic Indexing Method for Medical Documents. *Computers and Biomedical Research* 25: 336–350.

Waldhart, Thomas J. 1974. Utility of scientific research: The engineer's use of the products of science. *IEEE Transactions on Professional Communication* PC-17, 2 (1974): 33–35.

Walton, E. 1962. Motivation to communicate. *Personnel Administrative* 25(2): 17–19, 39.

Walton, Kenneth R. 1986. SearchMaster programmed for the end-user. *Online* (September): 70–79.

Wanger, Judith, Carlos A. Cuadra, and Mary Fishburn. 1976. *Impact of On-line Retrieval Services: A Survey of Users, 1974–75.* System Development Corporation.

Ward, Mark. 1991. Are you a team killer? *EDN (Electronic Design News)* 36(14A): 1–2.

Ward, Martin. 2001. A survey of engineers in their information world. *Journal of Librarianship and Information Science* 33(4): 168–176.

Watanabe, T. 1980. Visual communication technology: Priorities for the 1980s. *Telecommunications Policy* 4(4): 287–294.

Weggel, J. Richard. 1973. Coastal engineering IAC. In *Proceedings of the Meetings of Managers and Users of the DOD Information Analysis Centers,* pp. 125–140. Available from NTIS: AD/A008–289.

Weil, B. H. 1958. The role of the information service group in internal communications. In *Information and Communication Practice in Industry.* T. E. R. Singer, ed. Reinhold.

Weil, Ben H. 1977. *Benefits from Research Use of the Published Literature and the Exxon Research Center.* Washington, DC: National Information Conference and Exposition.

Weil, Ben H. 1980. Benefits from research use of the published literature at the Exxon Research Center. In *Special Librarianship: A New Reader.* Eugene B. Jackson, ed., pp. 581–594. Metuchen, NJ: Scarecrow Press.

Weinschel, Bruno O., Russel C. Jones, principal investigators. 1986. Toward the more effective utilization of American engineers: The National Engineering Utilization Survey. Washington, DC: American Association of Engineering Societies.

Weiss, R. S., and E. Jacobson. 1955. A method for the analysis of the structure of complex organizations. *American Sociological Review* 20: 661–668.

Welborn, Victoria. 1991. The cold fusion story: A case study illustrating the communication and information seeking behavior of scientists. *Science and Technology Libraries* 11(3): 51–58.

Wersig, G. 1973. Zur Systematik der Benutzerforschung. *Nachr. Dok.* 24(1): 10–14.

White, Howard D., and Katherine W. McCain. 1989. Bibliometrics. In *Annual Review of Information Science and Technology* 24. Amsterdam: Elsevier Science Publishers.

White, P. A. F. 1980. *Effective Management of Research and Development,* 2nd ed. MacMillan.

Whitley, Richard, and Penelope Frost. 1973. Task type and information transfer in a government research laboratory. *Human Relations* 25(4): 537–550.

Whitmire, Ethelene. 2002. Disciplinary differences and undergraduates' information-seeking behavior. *Journal of the American Society for Information Science and Technology* 53(8): 631–638.

Whittaker, John D., and Ted G. Eschenbach. 1998. Connecting what engineers do with how they are taught. In *Proceedings of the 1998 Annual ASEE Conference.* Washington, DC: ASEE.

Wilkinson, K. R., C. J. Finelli, E. Hynes, and B. Alzahabi. 2000. University-wide curriculum reform: Two processes to aid in decision making. In *30th Annual Frontiers in Education, Conference, Building on a Century of Progress in Engineering Education. Conference Proceedings,* Vol. 1, October 18–21. Kansas City, MO: IEEE.

Williams, Frederick, and Eloise Brackenridge. 1990. Transfer via telecommunications: Networking scientists and industry. In *Technology Transfer: A Communication Perspective,* Frederick Williams and David V. Gibson, eds., pp. 172–191. London: Sage Publications.

Williams, Frederick and David V. Gibson, eds. 1990. *Technology Transfer: A Communication Perspective.* London, UK: Sage Publications.

Williams, F. W., and J. M. Curtis. 1977. The use of on-line information retrieval services. *England: Program: News of Computers in Libraries* 11(1): 1–9.

Williams, Julia M. 2002. The engineering portfolio: Communication, reflection, and student learning outcomes assessment. *International Journal of Engineering Education* 18(2): 199–207.

Williams, Julia M. 2000. Transformations in technical communication pedagogy: Engineering, writing, and the ABET Engineering Criteria 2000. In *Technology & Teamwork,* pp. 75–79. IEEE.

Williams, Martha E. 1994. The Internet: Implications for the information industry and database providers. *Online and CD-ROM Review,* 18(3): 149–156.

Williams, Martha E. 1985. Electronic databases. *Science* 228 (April 26): 445–456.

Williams, Martha E. Annual since 1982. *Information Market Indicators Reports.* Monticello, IL: Information Market Indicators.

Williams, Martha E. 1975. *The Impact of Machine-Readable Data Bases on Library and Information Services.* Urbana, IL: University of Illinois Information Retrieval Research Lab.

Wilson, T. D. 1981a. On user studies and information needs. *Journal of Documentation* 37(1): 3–15.

Wilson, T. D. 1981b. *Guidelines for Developing and Implementing a National Plan for Training and Education in Information Use.* UNESCO, PG1/80/WS/28.

Wilson, T. D. 1983. Office automation and information services. In *Proceedings of 7th International Online Information Meeting,* December 6–8, 1983, London, England, pp. 199–206. Learned Information.

Wilson, T. D. 1981. On user studies and information needs. *Journal of Documentation* 37(1): 153–157.

Wilson, T. D., et al. 1979. Information needs in local authority social services departments: A second report on Project INISS. *Journal of Documentation* 35(2): 120–136.

Wilson, T. D., and D. R. Streatfield. 1977. Information needs in local authority social services departments: An interim report on Project INISS. *Journal of Documentation* 33(4): 277–293.

Wilson, Tom D. 1994. Information needs and uses: Fifty years of progress? In *Fifty years of Information Progress: A Journal of Documentation Review*, pp. 15–51. London: ASLIB.

Winsor, D. 1996. *Writing Like an Engineer, a Rhetorical Engineer.* Mahwah, NJ: Lawrence Erlbaum Associates.

Winsor, D. A. 1988. Communication failures contributing to the Challenger accident: An example for technical communications. *IEEE Transactions for Professional Communications* 31(3): 101–107.

Winsor, Dorothy A. 1990. How companies affect the writing of young engineers: Two case studies. *IEEE Transactions on Professional Communication* 33(3): 124–129.

Wiseman, Norman, Chris Rusbridge, and Stephen M. Griffin. 1999. The Joint NSF/JISC International Digital Libraries Initiative. *D-Lib Magazine* 5(6). Available from World Wide Web: <http://www.dlib.org/dlib/june99/06wiseman.html>.

Woelfle, R. M. 1979. Role of the engineer in improving the communication of technical information. *IEEE Transactions on Professional Communication* PC–22(1): 24–26.

Wolek, Francis W. 1969. The engineer: His work and needs for information. *Proceedings of the American Society for Information Science* 6: 471–476.

Wolek, Francis W. 1970. The complexity of messages in science and engineering: An influence on patterns of communication. In *Communication Among Scientists and Engineers.* Carnot E. Nelson and Donald K. Pollock, eds., pp. 233–265. Lexington, MA: Heath Lexington Books.

Wolek, Francis W. 1972. Preparation for interpersonal communication. *Journal of the Society for Information Science* 23(1): 3–10.

Wolek, Francis W. 1973. Professional work: The context for evaluating the impact of user studies. In *15th Annual National Information Retrieval Colloquium*, Philadelphia, May 3–4 1973, pp. 34–35.

Wood, D. N. 1971. User studies: Review 1966–1970. *ASLIB Proceedings* 21: 11–23.

Wood, D. N., and D. R. L. Hamilton. 1967. *The Information Requirements of Mechanical Engineers: Report of a Recent Survey.* Library Association.

Woodward, J. 1965. *Industrial Organization: Theory and Practice.* OUP.

Wooster, Harold. 1967. Policy planning for technical information in industry. In *Documentation Planning in Developing Countries, International Federation for Documentation (FID/DC) Symposium,* Bad Godesberg, Federal Republic of Germany, November 29, 1967, 16 pages.

Workshop Steering Group. Report of a workshop held in Washington, DC, June 11, 1985. Washington, DC: National Academy Press, 1985, 52 pages.

Wright, N. H. 1982. Matrix management: A primer for the administrative manager. *Engineering Management Review* 10(3): 65–68.

Yamada, H. 2001. Change management and the New Industrial Revolution. In *IEMC'01 Proceedings,* pp. 240–245. October 7–9. Albany, NY: IEEE.

Yatchisin, George, LeeAnne Kryder, Marty Williams, and Mark Kerr. 1998. Educating engineers to communicate in the 21st century: University of California, Santa Barbara's first year Engineering Communication Sequence. In *Proceedings of the 1998 Society for Technical Communication Annual Conference,* pp. 17–19. Arlington, VA: STC.

Yates, F. 1965. *Sampling Methods for Censuses and Surveys,* 3rd ed. Griffin.

Yates, J. K. 2001. Retention of nontraditional engineering and construction professionals. *Journal of Management in Engineering* 17(1): 41–48.

Yeaple, R. N. 1992. Why are small research-and-development organizations more productive? *IEEE Transactions on Engineering Management* 39(4): 332–346.

Young, J. F., and L. C. Harriott. 1979. The changing technical life of engineers. *Mechanical Engineering* 101(1): 20–24.

Zainab, Awang Ngah, and A. J. Meadows. 1999. Electronic support and research productivity: The case of academic engineers and scientists. *Malaysian Journal of Library and Information Science* 4(1): 71–85.

Zand, D. E. 1981. *Information, Organization, and Power: Effective Management in the Knowledge Society.* New York: McGraw-Hill.

Zhang, Yin. 2001. Scholarly use of Internet-based electronic resources. *Journal of the American Society for Information Science and Technology* 52(8): 628–654.

Zia, Lee L. 2000. The NSF National Science, Mathematics, Engineering, and Technology Education Digital Library (NSDL) Program. *D-Lib Magazine* 6(10). Available from World Wide Web: <http://www.dlib.org/dlib/october00/zia/10zia.html>.

Zielstorff, Rita D., Christopher Cimino, G. Octo Barnett, Laurie Hassan, and Dyan Ryan Blewett. 1993. Representation of nursing terminology in the UMLS Metathesaurus: A pilot study. In *Proceedings: 15th Annual Symposium on Computer Applications in Medical Care: Assessing the Value of Medical Informatics,* pp. 392–396.

Zimmerman, Donald E., and Michael Palmquist. 1993. Enhancing electrical engineering students' communication skills using online and hypertext aids. In *Proceedings of the 1993 IEEE International Professional Communication Conference,* pp. 428–431. Piscataway, NJ: IEEE.

Zimmerman, Donald E., Michael Palmquist, Kate Kiefer, David Vest, Marilee Long, Martha Tipton, and Laura Thomas. 1994. Enhancing electrical engineering students' communication skills: The baseline findings. In *Proceedings of the IEEE International Professional Communications Conference,* pp. 412–417. Piscataway, NJ: IEEE.

Zinn, Karl. 1976. A computer-based system to enhance sharing of technical information in a system of scientific communities. In *Current Research on Scientific and Technical Information Transfer.* Micropapers edition. New York: Jeffrey Norton Publishers, pp. 15–24.

Zipperer, Lorri. 1993. The creative professional and knowledge. *Special Libraries* 84(2): 69–78.

Zirnan, J. M. 1969. Information, communication, knowledge. *Nature* 224, 8 pages.

INDEX

AAU Research Libraries Project, *see* Association of American Universities Research Libraries Project

Abels, Eileen G., 52, 69

Accreditation Board for Engineering and Technology, 167
  Engineering Criteria (ABET EC2000), 99–100, 102–103

Ackoff, R.L., 12–13, 123

ACM, *see* Association for Computing Machinery

Adrezin, Ronald S., 105

Aerospace Engineering, 167, 174

Aerospace Information Management-UK (AIM-UK), 51, 67

Age, *see* Demographics, age

AIAA, *see* American Institute of Aeronautics and Astronautics

Allen, Thomas J.
  channels, 25, 31–32, 61, 62–63, 63–64
  communication patterns of engineers and scientists, 150
  factors influencing selection of sources, 35

influence of, 33

journals, use of, 119–120, 121, 123–124

research of, 4, 5

time spent on reading, 123

work environment, 151–152

Al-Shanbari, H., 73, 74, 75, 86–87, 94, 158

Amare, N., 84–85

American Institute of Aeronautics and Astronautics, 166, 167, 169–172, 174, 175, 177

American Institute of Physics, 130

American Society for Engineering Education, 185

American Society for Testing and Materials, 167

Anderson, Thad, 30, 75–76, 100–101, 107

Andrys, Christine, 87, 158

Arizona State University, 102

ARL Directory of Scholarly Electronic Journals and Academic Discussion Lists, 116

arXiv.org, *see* Los Alamos National Laboratory arXiv.org

Communication Patterns of Engineers. By Carol Tenopir and Donald W. King   **253**
ISBN 0-471-48492-X © 2004 Institute of Electrical and Electronics Engineers

ABOUT THE AUTHORS

Carol Tenopir is a professor at the School of Information Sciences at the University of Tennessee, Knoxville. Her areas of teaching and research include: information access and retrieval, electronic publishing, the information industry, online resources, and the impact of technology on reference librarians. She is the author of five books including, *Towards Electronic Journals: Realities for Scientists, Librarians and Publishers* (Special Libraries Association, 2000) with Donald W. King. Dr. Tenopir has published more than 200 journal articles, is a frequent speaker at professional conferences, and since 1983, has authored "Online Databases," a column for *Library Journal*. She is the recipient of the 1993 Outstanding Information Science Teacher Award from the American Society for Information Science/Institute for Scientific Information and the 2000 ALISE Award for Teaching Excellence. Dr. Tenopir also received the 2002 American Society for Information Science & Technology Research Award for lifetime achievement in research and holds a Ph.D. in Library and Information Science from the University of Illinois.

Donald W. King is currently a research professor at the University of Pittsburgh School of Information Sciences. Formerly, He was at King Research, Inc. and Westat, Inc., where his 40-year career focused on research and description of communication systems and services. Recent emphasis has addressed electronic journals and scientific communications through a series of studies with Carol Tenopir. He has co-authored or edited 17 books and hundreds of formal publications in this area. In recognition of his contribution to the field, he was named a Pioneer in Science Information by Chemical Heritage Foundation. He has earned the Pioneer in Information Science Research Award, and Award of Merit by the American Society for Information Science & Technology. He is a Fellow of the American Statistical Association, has received the Miles Conrad Award, and is an Honorary Fellow of the National Federation of Abstracting and Information Services.